Electronic Surveillan

This book is dedicated to all my family and friends, with special thanks to Neil, without whose help I would not have been able to tell my ASCII from my elbow.

Electronic Surveillance Devices

Second edition

Paul Brookes

Newnes

OXFORD AUCKLAND BOSTON JOHANNESBURG MELBOURNE NEW DELHI

Newnes
An imprint of Butterworth-Heinemann
Linacre House, Jordan Hill, Oxford OX2 8DP
225 Wildwood Avenue, Woburn, MA 01801-2041
A division of Reed Educational and Professional Publishing Ltd

 A member of the Reed Elsevier plc group

First published 1996
Reprinted 1997, 1998, 1999 (twice)
Second edition 2001
Reprinted 2002

British Library Cataloguing in Publication Data
Brookes, Paul
  Electronic Surveillance Devices
  I. Title
  621.38928

ISBN 0 7506 5199 7

For more information on all Butterworth-Heinemann
publications please visit our website at www.bh.com

Typeset by Avocet Typeset, Brill, Aylesbury, Bucks
Printed and bound in Great Britain by Biddles Ltd, *www.biddles.co.uk*

Contents

Preface

Since humans first communicated, the thirst for information has never been quenched. The importance, and value, of information can never be understated. There cannot be many people who have not wished, at some point in their lives, to be 'a fly on the wall', to know for certain what has been said, or what has taken place.

Electronic surveillance has been used for many years, the first case being the use of simple hard-wired microphones placed near the enemy trenches, to enable the listener to be aware of troop movements, and imminent attack. During the Cold War, the flow of anecdotes and rumours about the use of surveillance devices ranged from the sublime to the ridiculous. Inspired by these anecdotes, a new breed of electronics enthusiasts began designing their own devices, often having the designs published in enthusiast magazines. This phenomenon happened at the same time as a relatively new invention, the bipolar transistor, became available to the general public. Although very expensive, in short supply, with only a few device types available, the small transistor would soon replace the old fashioned, large, energy-hungry thermionic valve, now left on the shelf. With the transistor, which would work on a supply from a battery instead of bulky mains supply transformers, it was possible to build small circuits that could operate as amplifiers, switches and transmitters.

Armed with the new miniature electronic components, the designer was able to build smaller and smaller circuits. The race was now on, to build smaller and more efficient devices. Units, designed to act as audio transmitters, were disguised as (inedible) olives, with rather ambitious quotes for transmission range. Inspired by the large number of techno-spy movies, devices appeared in many other disguises and forms such as transmitters inside shoe heels, microphones in buckles and brooches, etc.

Soon afterwards, in certain countries, a cry was heard that these non-official surveillance devices were being used by many people or groups, whose aim was to obtain information that they were not supposed to have. This lead to a strict and severe clampdown on the manufacture, use and sale of devices specifically intended for monitoring or intercepting conversations. Governments introduced new privacy laws to help cope with the growing problem of unauthorized electronic intrusion. Some loopholes in legislation did appear at one point, with a few manufacturers hastily tearing off the labels of 'secret transmitters' and replacing them with 'baby monitor' stickers, re-naming 'automatically switching telephone to tape recorder' as 'telephone secretaries', etc.

Since this time, when miniature electronics was in its infancy, electronic surveillance has matured into big business, with applications for the numerous range of devices in many walks of life. This book is intended to educate and inform anyone who is involved in the security of premises, or the security and protection of others or themselves. Several chapters of this book describe the circuit diagrams of several devices that can be used for surveillance. These diagrams have been included as information both for the electronic engineer who is involved in the development of security devices, as well as to show security personnel the type of devices available, and how they function. To make the book even more helpful to security personnel, a complete chapter has been devoted to the topic of counter-surveillance devices and techniques. The last chapter has been included to make the reader think hard about security, and perhaps make some aware of the potential danger of giving information freely and accidentally. With regards to surveillance, although some people may cry out '1984', many people feel safer when parking in an area that has full video protection, or understand that they live better lives with the knowledge that electronic surveillance is being used as one of the tools against crime.

A note regarding this second edition of Electronic Surveillance Devices

The author of this book decided that a lot more material, concerning the practical side of constructing surveillance devices, could be included in this second edition. To this end, the section regarding microphones has been expanded so that it now includes such items as construction methods for the popular tube microphone, spike microphone, probe microphone and the pen microphone.

Other new topics include audio amplifiers, the use of lasers for voice interception, noise filters, relay transmitters, tracking systems and beacons, high powered transmitters, the 'look – no batteries' parasitic transmitter, 'Trojan horse' devices and power supplies, automatic camera video record switching devices, snooper detectors, telephone circuits, etc. with the occasional anecdote of the type that reflects the author's popular style, thrown in.

As was the case with the first edition of this book, this new edition will help all readers gain a better understanding of what electronic surveillance is all about, as well as saving a large sum of hard-earned cash by the reader building their own equipment. When compared to other electronic items for sale on the market, electronic surveillance devices contain relatively few components for such an astronomical mark-up price. A device that may cost very little to build can be professionally encased or put into a small plastic 'potting box' with some epoxy resin glue. The resulting unit, built in just two or three hours, can sometimes be retailed at the equivalent of one week's wages. One or two hours of fruitful electronic surveillance can also warrant similar payment. Several successful new companies that manufacture and sell their own surveillance devices and systems have been born since the first edition of *Electronic Surveillance Devices* was first published – maybe you will be the next.

1 Why use electronic surveillance?

Many individuals and organizations may, for various reasons, wish to use electronic surveillance techniques at some time or another.

First we will look at the possible needs of an individual whose use of surveillance devices may be as follows:

1 For security reasons, e.g. monitoring peripheral areas of their property.
2 In pursuit of hobbies, such as the remote monitoring of wildlife activities.
3 The requirements for checking on spouse loyalty, to check on the misuse of telephones, etc.

Security

In these days of rising crime rates, more and more people are becoming security minded. Video cameras can be installed to give a view of areas that are vulnerable to attack, and these can be radio-linked, using low powered video transmitters, to a control centre. At the other end of the scale, a simple audio transmitter can be used to monitor sounds made by intruders at the fraction of the cost of an expensive system.

Hobbies

For many years naturalists have been using electronic devices in pursuit of their hobby. Probably the most well-known of these units are the microphones used to listen to, and record, animal noises. The microphone may be of a simple sort that is supplied with a standard cassette recorder, or of a highly directional type that can pinpoint a noise source whilst cutting out any surrounding unwanted noises. Once again a humble audio transmitter can be placed near the vicinity of the area occupied by the animal, with a sound activated receiver and tape recorder combination situated inside the relative warmth of the observer's hide, or car, situated a few hundred metres away if so required.

Spouse loyalty

Unfortunately (or fortunately if you sell the necessary equipment) this is the largest reason for the sale of low-budget surveillance equipment. It was noted a few years ago that advertisements began to appear for 'room transmitters' in some of the well-known tabloids. Although some secret service agents must occasionally use these types of publications for wrapping up their kitchen waste, the longevity of such adverts was proof enough that members of the general public were buying these products. Not only 'room transmitters' were sold, but many 'telephone recorders' and 'telephone transmitters' were also offered alongside them. Telephone recorders are used to record both sides of a telephone conversation, recording not only the conversation, but if played through a decoder, the number dialled can also be revealed.

We will now consider the reasons why a business may wish to use electronic surveillance equipment. Information may need to be gathered on:

1 deals, tenders and meetings;
2 checking on colleagues and employees for dishonesty such as pilfering, disloyalty, double-crossing, etc.;
3 general security measures.

Unlike the personal reasons for using surveillance equipment, the reasons that an organization may require such equipment is by its very nature sensitive, and very rarely disclosed. Two case histories follow, which are typical examples of how electronic methods of 'information acquisition' may be employed.

Case 1

A self-employed tradesman was doing some renovation work in London. Whilst his team were working there, he was covering another job in York.

Every week, hundreds of pounds worth of material was going missing from his London storeroom, so he asked a 'security advisor' if it were possible to put a transmitter in the London storeroom that he could listen to on a radio in York. He was advised that such a system would cost a small fortune to set up, so another method would be devised.

Since the thefts were taking place between the times when the foreman left the premises (around 6pm) and when the main site was secured by a night watchman (around 8pm), the following cost effective solution was agreed.

The customer purchased a 24-hour timer socket, that was set to switch on for the two hours in which the pilfering was believed to have been taking place. A pressure mat, as used in a standard burglar alarm, was fixed under a nailed-down carpet at the doorway of the storeroom, then the wires were hidden away neatly and connected to a slow speed cassette recorder, with a good microphone amplifier and 20 minutes hold-on timer. The cassette recorder had been modified, with a motor speed control, to give around two hours of recording time on one side of the tape, ample for the purpose intended. On the Sunday, the system was fitted into the storeroom, tested, and then the employer made his way back to York.

The following Saturday, the employer drove back to London and rewound the tape. On playing the tape back, it was found that the system had been activated on three separate occasions, with voices of two of his employees discussing the choice of materials to be stolen.

On the Monday morning, he called his gang of eight workers into the office for a 'meeting', and once everyone had settled down, he started off the proceedings by saying how the pilfering of stock had reached such proportions that his firm was now considering sacking all the workers. This lead to a silence, which he broke by saying he just wanted to play a tape recording of something rather interesting. He pressed the play button and watched the faces of the two pilferers. After a token protest, strangely enough, three men walked out of the room!

How much did this system cost the employer? Considering the fact that the 'stock shrinkage' had cost around £2 500 to his knowledge at least, the cost of less than £100 seemed rather a bargain. It was also a good job that the thieves never stopped talking whilst in the storeroom!

Case 2

The use of, and the reason for, some types of surveillance equipment has sometimes got to be questioned on moral grounds. A self-employed person, 'Mister Smith', who prided himself on his standard of workmanship and what he believed to be very fair prices indeed, found that his work was dropping off. His business relied on Mister Smith obtaining contracts for his work. He was strictly a one man outfit, and always considered his bid to be far lower than his larger competitors.

Mister Smith really believed himself to be the cream of his profession (that of painting and artexing), so why was he losing out when trying to get some contracts? Deciding that there was something going on that was not 'fair play', he began to think of what he might do to get to the bottom of what he thought was probably a case of corruption.

The method used by one particular contractor was to sit inside his portable cabin office and interview each representative on the shortlist in turn regarding their bid. Mister Smith decided that this was the best and only chance of finding out the truth of the matter.

He had obtained a cheap 'disposable' micro transmitter that could transmit voices in a room to a standard VHF radio up to 100 m away to be received on a standard VHF radio. He managed to get into the portable cabin when it was empty, prior to the first interview being held, then by using adhesive tape, fastened the transmitter under the desktop of the contractor.

The disgruntled gentleman was the second of the four on the shortlist to be interviewed, so after his turn, he went back to his car (which was parked around the corner from the office), and tuned his car radio onto the transmission coming from his hidden transmitter. Sure enough, he could listen to every breath being drawn in the office, even managing to pick up a snippet of information that suggested to him that his price really was the lowest submitted. After the last person on the shortlist had been interviewed, the contractor and his colleague sat around discussing who was to be offered the work.

It came as a shock to the tradesman when he heard of himself being mentioned as a 'good bloke but too old to be considered taking on'. He was seventy-five after all was said and done. He retired and has no further use of surveillance devices.

There is a question of morality in the above, and certainly a moral, 'Those who eavesdrop never hear good of themselves'.

Security organizations

It is known that law enforcement agencies, investigative journalists and other organizations use surveillance devices in the course of their work. The innovation of CCD (charge-couple device) and pinhole cameras that can be hidden just about anywhere have opened new doors for evidence gathering and closed a few doors on those at the receiving end. The amount of telephone tapping being done by various security agencies with government approval can never be estimated, since the figures are seldom published.

How many 'micro room transmitters', 'telephone recorders' and 'telephone transmitters' have been sold? The estimate runs into tens of thousands.

At this point it would be prudent to separate the image that is conjured up in the abundance of action packed, spy thriller films, from that of the true potential of transmitting devices currently available in terms of:

1 the range (distance) from which a surveillance transmitter can be received, and
2 the size and physical appearance of a miniature transmitter.

It does not matter if the transmitter is described or advertised as a miniature transmitter or a micro transmitter, you will not find a device the size of a shirt button in the envelope when it arrives through your letterbox. Even more important is that the range of transmission expected by some members of the general public (who are exposed to both spy films and certain 'over-enthusiastic' advertising sales brochures) is not accurate. To sum up the present situation, if you are hoping for a shirt button transmitter with a transmission range sufficient to beam your messages around the globe, forget it for the time being.

Surveillance transmitters, by their very nature, are small and do not generally have the room for big batteries. Not only that, but the whole point of covert surveillance transmitters is that they have just enough power to transmit information to the operative, or recipient, without being so strong that the 'secret' signal is picked up by every scanning receiver owner for a 100 mile radius.

A typical requirement for transmission distance would be between 25 to 500 m, although some companies have occasionally produced relay stations. These relay stations are simply a receiver and a transmitter wired together, with both items working on a different frequency. The receiver is tuned to the frequency of the 'planted' low power device, then the information is re-transmitted by the relay transmitter on a different transmission frequency to that of the original, in just the same fashion that standard television and radio broadcast relay transmitters do. Although making the system insecure, liable to be accidentally intercepted, the powerful relay station could be situated in a parked van, re-transmitting the weak transmissions from a surveillance transmitter to a distance of 20–30 miles away on a VHF frequency, or around the globe if the receiver is patched through (interfaced to) a mobile phone.

Whilst on the subject of 'accidental interception', it should be noted that there are now some cheap scrambling devices and circuits available, but their size makes them largely impractical for use in scrambling the audio of a miniature transmitter.

As can be seen from the above scenarios, electronic surveillance costs as much or as little as you wish, matching the importance of the operation. The price of equipment varies from the equivalent of a bottle of Scotch to an astronomical amount, but how valuable is information?

2   Types of devices

Electronic surveillance devices are categorized into the following groups, however it should be noted that there are some devices that are hybrids of two or more of each section.

Room transmitters

Room transmitters are, as their name implies, designed to be surreptitiously placed within an area of a building, thereby enabling the listener to hear any conversation (or noises) that might occur within that area. Depending upon the design of these transmitters, it is literally possible to hear a pin drop.

Leaving aside the various minor permutations of room transmitters for a while, the two main categories of these devices are:

1 battery powered
2 mains supply powered.

Battery powered room transmitters

Battery powered room transmitters are devices that are powered by internally contained batteries or by means of a clip-on type of battery. The clip-on battery type of unit is usually found in the cheaper range of transmitters or when a longer period of surveillance, or longer range of transmission, is required.

Battery powered transmitters can be produced or hidden in a vast range of disguises that include:

 1 inside calculators
 2 inside pens
 3 inside or behind clocks
 4 behind picture and photograph frames
 5 inside plywood doors
 6 underneath carpets
 7 inside clothing
 8 behind curtains
 9 underneath or inside furniture
10 inside briefcases
11 left in wastepaper baskets.

Obviously, the list of places to conceal transmitters or finding articles to disguise them as, is endless, which is one of the reasons that battery powered transmitters have held a fascination for many people for a good many years!

The type of battery used in conjunction with these transmitters will vary, depending on the design of the transmitter and the requirements of the installer. Many of the self-contained transmitters (those units with a built-in battery compartment) use one or more of the small watch batteries, or better still, the hearing aid type of battery, the latter having a better performance. If size does not present a problem regarding the concealment of a transmitter, a hefty PP3 style 9 V battery, or even larger type, can be used. The larger the battery, the larger the storage capacity, allowing a longer transmission time.

The size and shape of a miniature transmitter will vary, but the average self-contained unit is around 19 mm × 12 mm × 9 mm deep, indeed, a model suitable for clipping onto a PP3 style battery would be the same size as the battery, since there is no point in making a transmitter any smaller if it needs such a large battery!

If a transmitter is built within a pen, the unit will not be able to have a reasonable aerial, unless the unit is worn on the body and an aerial wire is concealed on the person. The transmission range of a pen transmitter will only be around 15 or 20 m, but can be increased if powerful lithium batteries are incorporated into the design.

Transmitters built into calculators tend to have a similar transmission range to pen transmitters but, depending on the size of the calculator, it is possible to get a reasonable length of aerial wire installed inside the calculator casing.

It is an impressive point to note that the assortment of disguises for room transmitters are still fully functional, i.e. the pen transmitters still write and calculator transmitters still calculate.

Mains powered room transmitters

Mains powered room transmitters are powered by a mains voltage supply, with some models designed to provide constant trickle charging of nickel cadmium (NiCad) batteries to cover the possibility of the mains power being interrupted for any reason.

The mains powered transmitter needs some means of dropping the mains voltage level of around 240 V, alternating current, to a direct current, low voltage of around 6–18 V. This means that the device has to employ not only the transmitter circuitry but also the required voltage dropper, rectification, smoothing and voltage stabilization circuits. This may seem a lot of circuit to cram inside a two-way plug-in mains adaptor, but it can be, and is, done.

If there is sufficient room, a standard mains to low voltage supply that utilizes a step-down transformer can be used, which is far more reliable than capacitor leakage or dropper resistor techniques. If a device uses a transformer, although an audible buzz can be produced from the unit into the room (caused by a poorly constructed trans-former with vibrating laminations), the unit can supply a consid-erable current to a high-powered transmitter. If a mains dropper resistor circuit is used, unless only a small current is drawn through the resistor, the power, in the form of heat, can be high. The most popular method for supplying the small current and voltage require-ments of a typical transmitter is by the use of a high voltage capacitor, which acts as a resistance at the 50 Hz mains supply frequency. The only drawbacks with this device are that poor quality capacitors have the habit of breaking down and the current allowed through them is relatively small, so they are not able to drive a powerful transmitter. This problem can be overcome by allowing a small amount of radio power produced by the circuit to be injected into the mains wiring, but this does add more complexity to the circuit.

Mains powered transmitters can be hidden in almost any place where there is mains power available, or built into any mains powered appliance such as:

 1 inside walls and partitions
 2 inside ceilings
 3 under floorboards
 4 behind skirting boards
 5 inside burglar alarm control boxes
 6 built inside multi-outlet adaptors
 7 built into standard mains socket outlets
 8 inside office equipment
 9 inside mains powered wall clocks
10 inside table lamps, lamp holders, etc.

As for battery powered room transmitters, the list of places for possible concealment and disguise of mains powered transmitters is as long as the imagination. Because the plug-in double adaptors are very quickly installed, they rank amongst the favourite device of this classification. However, if a double outlet adaptor were to suddenly appear, i.e. one was not there before, they may also disappear just as quickly. (You cannot trust anybody these days!)

If the mains powered device is of the swapover mains outlet socket type, the installation will require isolation from the mains, but if the installer is competent, and no damage to wall coverings is caused, this can take only a few minutes. Suppliers of socket outlets with built-in transmitters will often be very obliging and supply the operator with an identical swapover unit.

Stand-alone mains powered transmitters are supplied in various housings, typically a square plastic box measuring 50 mm × 50 mm × 18 mm deep, or as an encapsulated board. Two trailing leads come from these units, intended for connection to the 'live' and 'neutral' lines of a domestic mains supply. These units can then be hidden inside walls, ceilings, mains powered office or domestic equipment, etc.

As mentioned previously, it is possible to design a mains powered transmitter with a built-in trickle charger that can keep rechargeable batteries topped up. This system is required if it is suspected at any time that the mains supply could be switched off by the target, these backup batteries can then keep the transmitter operational without a break in transmission.

Consideration of battery versus mains powered devices

The different advantages and disadvantages of the two different types of transmitter are given below.

Mains powered transmitters

Against

1 All mains powered transmitters, with the exception of a plug-in double adaptor type, will require, for obvious safety reasons, the isolation of the supply before installation commences.
2 Installation may involve a great deal of digging out holes in walls and associated remedial work.
3 Once installed, it may be impossible to regain entry to the target area to replace a broken unit.
4 If the device is hidden inside a portable disguise, there is a possibility of the unit being moved to a totally different area, e.g. pieces of office machinery are often moved around between areas.
5 Once installed, the chances are that the device cannot be removed quickly if things are getting 'out of hand', for example, someone has inadvertently picked up the transmission, and everyone in the office block spends their lunch-hour with earphones plugged in.

For

1 A good mains powered design will last indefinitely.
2 The unit will not require the regular replacement of batteries.
3 Depending upon the design, it is possible to obtain as much power as required for the transmitter.
4 If hidden well, without any physical evidence of installation, within the structure of a building, and if subsequently pinpointed by a detector, the finder would be wary of trying to dig it out (depending, of course, on his fee, but it would go on his report as suspect).

Battery powered transmitters

Against

1 Because the life of the battery (or batteries) that powers the transmitter is limited, there will be a need to have access to replace them regularly.
2 Due to the design of a battery powered device, to conserve battery life, the current drawn from the battery has to be kept to a minimum, therefore the radiated power from such a device will be low, so the range of the device will be somewhat limited.

For

1 The installation of a battery powered transmitter is relatively easy, as they are 'ready to go' without having to interrupt the mains supply.
2 If only a short time (up to around 80 hours) of information gathering is required, the problem of battery replacement does not then exist.
3 So called 'disposable transmitters' can be quite cheap compared to mains powered devices.

Telephone transmitters

As their name implies, these units are intended to be connected, by various methods, to the target telephone system and transmit information to a nearby receiving station. There are two basic models of telephone transmitter which are:

1 the series connected transmitter;
2 the parallel connected transmitter.

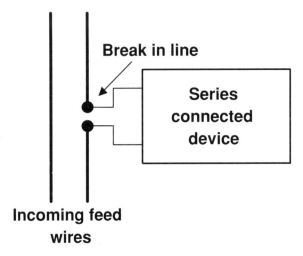

Figure 2.1a   *Method for series connection*

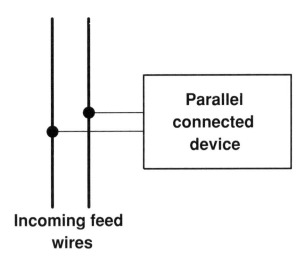

Figure 2.1b   *Method for parallel connection*

The methods of series and parallel connection are shown in Figure 2.1a and Figure 2.1b respectively.

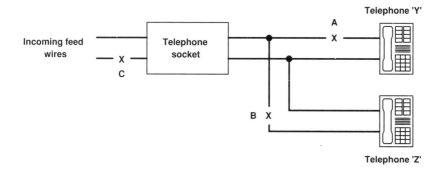

Figure 2.2 *Series connection in a multiple telephone system*

Series connected transmitters

Figure 2.2 shows a standard telephone system with one telephone extension. If a series transmitting device were to be connected at point A in the system line, then transmission of information would only take place if telephone 'Y' were used. Likewise, if a series transmitter were connected on the line at point B, then only information from telephone 'Z' would be transmitted. However, if a series device were to be connected at point C, then information would be transmitted whenever either telephone 'Y' or 'Z' were used.

Note that if a multiplex telephone system is used in the building, or if the phone company have installed a similar system (where more than one signal is multiplexed onto one line pair), this would mean that the operative would have to use a demultiplexer or install a transmitter on each individual telephone.

Parallel connected transmitters

Referring to Figure 2.2, it can be seen that the phone line pair, telephone 'Y' pair and telephone 'Z' pair, are all effectively in parallel with each other. From Figure 2.3, it can be seen that a parallel transmitter can be connected to both wires of the pair anywhere in the system, and, electrically, still be in the same place. Whenever either telephone is used, the information will be transmitted.

Why have series and parallel devices?

As was seen in Figure 2.1a and b, the series device has to be connected in series with the telephone line. This means that the actual wiring of the telephone system has to be physically

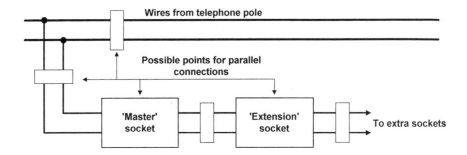

Figure 2.3   *Possible parallel connection points in a telephone system*

disconnected to allow the transmitter to be inserted. The installation of a series device does not necessarily mean that one of the telephone line wires has to be cut, the new ends stripped of insulation, and the device connected to these. This is because the transmitting device can be installed inside any junction box where there is sufficient physical space, or inside a telephone, utilizing any screw type connections that may be available.

Since many counter surveillance devices are now on the market that will give security personnel a warning that a line has been temporarily disconnected, simple checks (see Chapter 7) can detect some series devices. Also, as many lines are connected to intruder alarm systems, sounding a warning to a local law enforcement office, some operatives look towards the parallel device as a better option. Even if the series device is connected before the line is cut, a sensitive detector would note the sudden introduction of resistance in the line.

Unlike the series devices, a parallel connected device has the advantage that it can be installed without any temporary severance of a telephone line, and without the introduction of resistance. However, if the unit is parasitic, the voltage drop may be detected. The parallel device has another advantage in that it can be removed quickly, without causing much upset on the line voltage. Parallel devices may be supplied with crocodile style clips, which means the requirements for connection are a few millimetres of insulation-stripped telephone line or any suitable pair of terminals.

Some designs of series and parallel connected telephone transmitters are able to obtain their operating voltage from the telephone line itself, so are called 'leeches', 'suckers' or 'parasitic' devices. Other designs have their own battery, which in some cases are trickle charged from the telephone line.

The advantage of a parasitic device is that once installed, it does not need to be accessed in order to renew any batteries. It does have one drawback however – it has to draw current from the telephone line, making it easier to detect with measurements on the line. If the designer is too greedy for power, in the search for a longer transmission range, the circuit will not only be easy to spot by checks but, in the worst case scenario, cause the telephone line to be dragged down to an 'permanent' off-hook condition, and therefore to quick discovery. The transmission range of such a device is limited by the small current it dares to obtain from the telephone line, and so is typically 25–50 m, but some telephone transmitters may far exceed this figure by using extensive RF decoupling and then using the actual telephone line as an untuned long-wire aerial.

The battery powered telephone transmitter, unless of the trickle-charge type, will require constant access to facilitate battery renewal, and it will generally be designed as a parallel device. The two advantages of the battery powered type, however, are that since the device can be supplied with a greater operating current, the transmission range can be longer than that of the parasitic type, typically 500–1 000 m. Its second advantage is that since it only 'steals' a minute amount of audio from the telephone line, the possibility of discovery by electronic testing is greatly reduced.

Small, encapsulated and rainproofed telephone transmitters may be connected at any point of the exterior telephone wiring, from where the wires leave the building and even up to the telephone pole connections. Inside the building, a transmitter can be connected to the telephone line, hidden inside a telephone handset, or inside a telephone master or extension socket. Some firms have provided 'direct-swap' telephone socket units, where the standard printed circuit board inside the telephone socket has been replaced with one of a similar nature, i.e. surge protector, 'out of service' resistor, bell capacitor, transmitter.

The so called 'drop-in' telephone transmitter as seen in many spy films is designed to be installed as follows. An identical telephone microphone (mouthpiece) cartridge is rebuilt, so that it incorporates a transmitter. The telephone microphone of the target is removed, then the transmitting unit is installed. Although described as 'drop-in', the unit is usually attached with spade terminals, nuts, screws, etc. These units were most popular when every telephone handset was supplied by one company, and by implication, all telephones had the same microphone cartridge. With users now able to supply their own individual telephone handsets, the possibility of swapping over inserts has all but vanished.

As mentioned previously, there are allowable permutations of the above, with the availability of room and telephone transmitters combined in one package, utilizing the telephone current, and giving a formidable security risk.

Telephone line coupling

Sometimes it may be found that a simpler approach to the electronic acquisition of information on a telephone line is used. To acquire information, all that is required is some means of taking an audio signal off the line. This may be achieved by means that include:

1 Using an induction coil placed adjacent to the telephone line or to the handset itself.

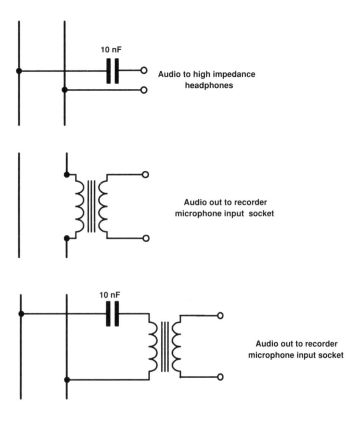

Figure 2.4   *Methods of telephone line coupling*

2 Attaching a pair of high impedance headphones via a capacitor of approximately 0.01 μF, across the telephone line pair.

3 Installing an audio transformer in series with one of the lines, with a tape recorder microphone input connected across the secondary winding of the transformer. A typical transformer would have a primary coil impedance of 8 ohms in series with the telephone line, with a secondary coil impedance to match the input impedance of the recorder's input, e.g. 500 ohms.

4 A combination of (2) and (3) above, which would require the coil attached to the telephone line side to have an impedance of around 2 kohms so as to not load the line.

Examples of the above can be found in Figure 2.4.

All of the above methods have various problems. An induction coil may take the form of the windings salvaged from an old relay or solenoid, but a unit known as a telephone pickup coil is offered for sale by many retail outlets. This is a coil of several hundred windings inside a plastic case, with a suction cup attached. Although advertised to be stuck onto the back of a telephone handset, behind the earpiece, in the older two-piece telephone, a telephone pickup coil will function if placed underneath the desk section of the telephone. The pickup coil comes with a flying lead, terminated with a standard 3.5 mm mono jackplug, intended to be plugged directly into the microphone socket of a cassette recorder. With reference to the 'spare' wire, often found in a telephone system (Figure 2.5), this is sometimes used in conjunction with one of the line pair to act as a remote control wire linkage to operate an off-hook condition, allowing the telephone microphone to relay conversations close to it.

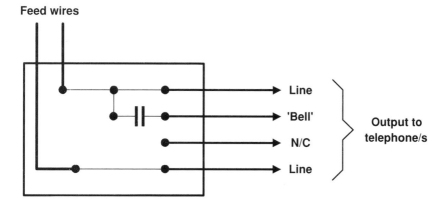

Figure 2.5 *Standard four-wire telephone connection*

Tape recorders

Any relatively low impedance, as presented by the circuits mentioned above, would be noted by counter surveillance techniques. Methods of telephone coupling, in their basic form, produce one problem – the operative would have to sit in an area, wearing headphones, for several hours a day, taking shorthand notes of all information gained. To avoid this problem, the information is fed to a tape recorder that can be slowed down to give extended recording time or activated by:

1 Voice activation (voice operated);
2 Line voltage activation.

Tape recorders are therefore available as standard or voice activated. Voice activated tape recorders are usually operated by switching to that mode of operation, then if a sufficiently loud noise is picked up by their in-built microphone, the recording mechanism begins to work, cutting off after half a second if no further noises are picked up, so as to conserve tape and battery life. For the operative, these units are ideal either as a stand-alone device, for monitoring standard room conversations, or with a small circuit, ideal for use as an automatic 'telephone recorder'.

A voice operated automatic telephone recorder is connected to the telephone system, where it will automatically begin to record whenever it picks up conversations on the line, but will stop recording if the line is either not in use or if there is a silence of more than half a second. The other design available, which is line voltage operated as opposed to voice activated, uses a standard cassette recorder combined with a voltage sensing switch circuit, which will activate the recorder if the voltage on the telephone line pair falls to a predetermined amount. This utilizes the fact that the line voltage drops considerably whenever the telephone handset is 'off the hook', with a corresponding rise when the handset is placed back on it.

Because a recorder will operate as soon as a number is dialled out, by playing back the number pulses, preferably slowed down, it is possible to ascertain what number has been dialled. If a tone dialler has been used, it is then necessary to use a more complex tone to number decoder. Due to the inherent amount of distortion produced by a slow, or extended recording time surveillance tape recorder, any tones recorded will be useless.

The 'infinity transmitter'

The invention of the infinity transmitter was conceived several years ago. In theory this device can be used to monitor room conversations anywhere in the world. Sounds too good to be true? The name in itself is somewhat misleading since it is not a transmitter as such, i.e. it is not a radio transmitter, but uses the telephone line as a transmission medium. The infinity transmitter is a device that is connected to the target telephone line. Inside the unit there is a tone decoder, switching circuits, high-gain voice amplifier, and a modulator circuit for imposing the audio (room conversations) onto the telephone line. The operation proceeds as follows.

As stated, the unit is connected to the telephone line of the target, in such a position as to enable the built-in microphone to pick up any conversations within range, e.g. up to around 10 m. The operative will then dial the telephone number of the target, and whilst dialling the last digit of the number, they will blow a 'special' whistle down their own telephone mouthpiece. Since the infinity transmitter has a tone decoder switch, which is tuned to the same frequency as the operative's whistle, the target telephone line is answered by the unit. The microphone in the unit will then pick up conversations within range, and send these down the telephone line to be listened to by the operative. Before the days of digital exchanges and tone dialling, these units were capable of 'answering' the telephone before the target telephone rang even once. Several drawbacks now exist with these ingenious devices – new systems tend to ring before the infinity transmitter can answer, raising the suspicions of the target. If the target area is served by a switchboard operator, the enable signal is frequently not allowed through before the switchboard operator orally announces the incoming call, and once again, a constant stream of silent callers, or 'sorry wrong numbers', or whistles being sounded, raises suspicion. Although there are several good tone decoder circuits available, and the old tin whistle can be replaced by a tone pad device, there are so many tones being used nowadays on the telephone line system for dialling, fax, e-mail, etc. that the days of the infinity transmitter seem to be numbered.

A device that has appeared on the market in the last few years is a variation of the infinity transmitter, and is much more sophisticated as can be seen in Figure 2.6. This device may be in a variety of permutations but operation is basically as follows.

Audio information is picked up by either a standard microphone or telephone tapping method. The audio information is then either hard-wired to a tape recorder or radio linked to a receiver and tape recorder combination. The tape recording may be played back, rewound or

Telephone line

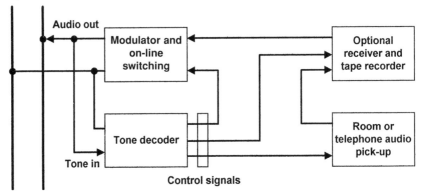

Figure 2.6   *The so-called 'infinity transmitter'*

forwarded (preferably downloaded on the operative's own telephone line in another office), to the operative who may be situated anywhere in the world, at leisure. This last part of the system is very similar in action to a standard remote access to an answering machine.

Even more complex systems, based on the above, have been designed. A control unit near the target area can be remotely accessed by a touch-tone pad to switch between microphones in monitored areas, to connect or disconnect any taps attached to the target telephone line, or to turn on or off any hidden transmitters in a building. The possibilities of remotely controlling devices, with the use of a dedicated telephone line (for use by the operative only), are endless. Even if the target line has to be used to download tape recordings, the downloading may be done at a fast speed, then played back at a normal rate. This not only provides a more secure system (if the tape plays whilst the target is about to make a telephone call then they will not be impressed at hearing their previous conversation), but also saves on the cost of the telephone call.

These infinity devices have often been supplied for security reasons, so that if a property is uninhabited, any security staff may telephone the office or other protected area so as to enable them to monitor for sounds made by any intruder, within the range of any microphone connected to the device. By using several microphones wired into the system, security control can have a voice/sound activated alarm which covers a large protected area, but be monitored from anywhere in the world via the telephone system.

Voice operated switches

Voice operated switches, abbreviated to VOX or VOX switches, find several applications in electronic surveillance. The VOX is a circuit that will pick up a sound, e.g. a conversation, and will then make a normally open connection close. Depending on the size of switch it contains, it will be capable of turning anything on (and off), from small low-power transmitters using just a few milliamperes of current, up to mains-powered security lights requiring several amperes of current. Although there are several tape recorders available on the market today that incorporate switchable VOX as part of their design, a VOX unit may be useful as an add-on for surveillance devices, such as voice activated transmitters or for the switching on of a CCTV (closed circuit television) recording system. If a VOX is used with a battery powered transmitter, although a negligible standby current is used, not only is battery life conserved somewhat, but the risk of RF detection, or interference, is minimized.

Microphones

Many microphones are available on the market but are generally divided into two groups:

1 omnidirectional;
2 unidirectional.

Omnidirectional microphones are designed to pick up audio signals in all, or any, direction. In practice, the pickup pattern will have preference to the front of the microphone, with an almost 'blind spot' at the rear. However, in surveillance systems, since the microphone is mounted in one fashion or another, and usually fixed in a set location, this is rarely a problem.

Omnidirectional microphone inserts come in all sizes, from a bulky crystal or dynamic type, measuring 25 mm in diameter, to the usual type which are often used in surveillance devices, electret microphone inserts that measure only 6 mm in diameter.

Unidirectional microphones are designed to be portable and capable of being pointed at the target. These are obtainable as parabolic dish microphones (either hand-held or tripod mounted) or rifle microphones. The parabolic dish microphone, which somewhat resembles a satellite dish, can have excellent results if the distance between the operative and the target conversation exceeds that at which a rifle microphone can be used. Depending on the gain of the dish and associated microphone and circuitry, good quality sounds may be obtained from a 45 cm diameter reflector over 250 m, with a fourfold increase in distance if the size of the reflector is doubled. Although extremely sensitive, the dish microphone would need to be fixed to a tripod, with aiming sights, to get the results mentioned previously. Large and cumbersome, although extremely lightweight if fashioned from plastic, glass fibre or carbon fibre, dishes are hard to use when there is anything more than a faint summer breeze. Also unless made from a transparent material, a parabolic microphone dish is very visible.

'Spike mikes' are microphones mounted at the end of a spike or probe. The microphone can be connected to the audio input of a miniature transmitter, to allow remote monitoring of conversations, to a headphone amplifier, or to a tape recording system for information retrieval at a later date. Some spike mikes and other 'listen through walls' devices do not use a standard microphone, but instead use a piezoelectric transducer microphone. These work on the same principle as the pickup needle of a standard record player. Although a needle picks up the vibrations caused by being squeezed through the groove of a record, the sounds of any partygoer's shouts are not picked up by the needle and amplified through to the loudspeakers! In the same fashion, a contact microphone used in a through-the-wall device has to be in contact with the target wall, which acts as a microphone diaphragm. Similar transducers are sometimes sold in music shops, as they can be used for harmonica pickups, or glued to the body of acoustic guitars.

Miniature electret microphone inserts

Electret microphone inserts are the most common type of microphone found in use in the vast majority of surveillance devices. These devices usually come in two types of package: the two-pin device is commonly 6 mm in diameter and 5 mm thick, or the larger version which is 10mm in diameter and 7 mm thick, whereas the three-pin device is supplied in the latter, larger package. The frequency response covers the range 50 Hz to 13 kHz, giving a good quality of transmitted sound. Both the smaller and larger devices have a supply voltage range of 1.5 V to 10 V.

Both the two-pin and three-pin devices are shown in Figure 2.7, along with the methods of connection for each device illustrated in Figure 2.8. As can be seen, the three-pin device has the advantage that both the resistor and capacitor are not required when connecting up to a circuit, therefore the inconvenience of having to solder on an extra connection leg to the insert is far outweighed by

Two pin

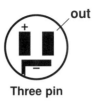

Three pin

Figure 2.7   *Miniature electret microphones*

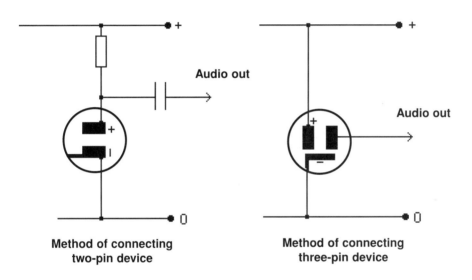

**Method of connecting
two-pin device**

**Method of connecting
three-pin device**

Figure 2.8   *Connection method of two-pin and three-pin electret microphone inserts*

saving component count when constructing a microtransmitter. When soldering legs to an electret insert which has not been supplied with legs by the manufacturer, great care must be taken to ensure that heat damage is not caused by taking too long to get the solder to run sufficiently. Inside the insert is a FET which can be quickly destroyed by excessive heat. The possibility of heat damage may be reduced if the insert is gently held in a metal notepad clip, and ensure that the wire to be soldered onto the pads is pre-tinned with solder initially.

Tube microphone design

The tube microphone should be included in every operatives' 'box of tricks', and is illustrated in Figure 2.9. The tube microphone can be poked through any hole that is available, such as a keyhole (or if not available, then a hole can be made using a drill bit, knitting needle, bradawl, hammer and chisel, etc.), so that the open end of the tube is actually in the same room as the target, but not sticking out of the hole. The other end of the hollow tube is glued carefully onto the aperture of a suitable microphone, with the end assembly potted in resin so as to prevent any noise from the outside of the room interfering with the conversation being monitored. One particular method of poking a tube into the target room, favoured by many operatives, is to use the same holes that are used for electrical socket cable or telephone socket cable. These services are very often wired up in a back-to-back fashion on the joining walls of offices and hotel rooms, as often are plumbing services, but these tend to be rather too noisy at times, due to the sound of running water being carried throughout the building by the pipework.

To prevent the risk of electric shock, both to the operative and to the target who may feel a great desire to start pulling at a strange tube that has suddenly appeared out of their wall, some measures must be taken. The microphone tube can either be fabricated from some suitable plastic, which tends to bend if pushed against an obstruction, or made from aluminium or steel tube that is covered in electrical insulation tape, or better still, heatshrink tubing. The soft cover will eliminate most insertion noise as well as act as an insulator against

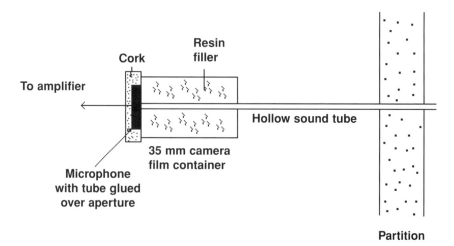

Figure 2.9   *Tube microphone design*

electric shock. Imagine what would happen if the target were to attach the tube to the mains 'live' supply, whilst the operative is holding the other end of the tube. The design shown uses a plastic 35 mm camera film container as a handy casing.

To judge the depth of penetration of the tube is somewhat of a hit and miss affair, but if the hole is being drilled, then a depth gauge could be employed, feeling very carefully for the drill completely penetrating the wall. The tube does not have to penetrate the adjacent wall, however, to obtain good copy.

It will probably be found that the most suitable microphone that can be used for fabricating a tube microphone assembly is an electret microphone insert, which, once the thin cloth cover is removed from the face of the microphone, exposes a small central hole that is ideal for gluing to a tube using epoxy resin adhesive. The diameter of the tubing chosen to act as the probe can be just a few millimetres if the sound is loud enough and the overall gain of the audio amplifier is high enough, but any larger size diameter could be used if the listening hole can accommodate it. Care must be taken during the gluing process so as to ensure that no adhesive spreads over the microphone hole, or gets inside the tubing.

When in use, it may be found that too much noise is generated by holding the tube microphone against the wall by hand, or that adjoining cable to the amplifier circuit rubs across the surface of the wall when disturbed. If this problem seems imminent, then the unit and cable can be taped to the wall so as to prevent all movement. An alternative approach to this design would be to connect a transmitter to the microphone assembly, as illustrated in Figure 2.10, so that it can

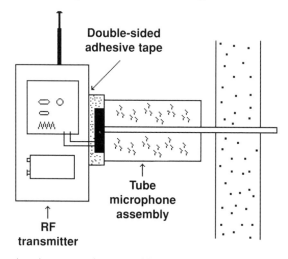

Figure 2.10 *Tube microphone transmitter assembly*

be left *in situ*, with conversations then being monitored either in the room, or at several hundred metres away if rapid response to a situation is not a concern. Any intrusive microphone assembly can be connected to a radio transmitter in a similar fashion.

The 'spike-mike' contact microphone construction method

The contact microphone construction is far more easier than fitting the microphone in place. An outer tube is fitted through a hole that has been drilled through a wall. The thinner steel or aluminium probe is then inserted into the outer tube, with preferably a piece of foam sandwiched between the two. This foam will act as a sound insulator so that sound being carried by the outer wall will not interfere with the targeted conversation. This is illustrated in Figure 2.11. It is now necessary to position the internal probe so that the point of the probe is making good contact with the plasterboard of the inner wall, which will vibrate in sympathy with sounds inside the room. The microphone itself is a contact type, but other designs may be experimented with. As was the case with the tube microphone assembly, it is possible to connect the output of the microphone device to the input of a transmitter so as to obtain a remote monitoring facility, although an audio amplifier or tape recorder can be used.

Probe-type microphone construction

The probe microphone seen in Figure 2.12 can be constructed in several different ways, depending upon the requirement of the operative. Basically, the probe microphone is constructed from a 6 mm diameter electret microphone insert, which is connected to a suitable length of thin screened cable. A length of aluminium tubing is then obtained, and one end is 'belled' out by a tool such as a dart, so that the microphone insert can be accommodated neatly inside the end of the tube if required. The other end of the tube must be reamed so as to prevent sharp edges cutting into the cable. A piece of heatshrink tubing is slipped over the cable and cable exit point of the tubing, and shrunk so as to give a nice finish to the microphone as well as some anchorage to prevent the cable being pulled and broken. A suitable jack plug is then connected to the end of the cable so that the microphone can be plugged into a variety of equipment. The whole of the tube can be covered in matt black heatshrink tube if so required so as to make probing more quiet, and to prevent the risk of a shiny metal tube catching the eye of the target. The probe microphone can be inserted into ready-made holes, such as keyholes

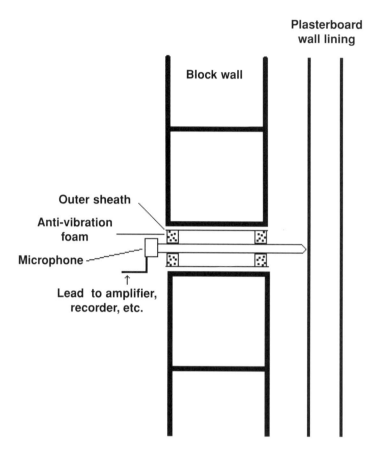

Figure 2.11   *'Spike' contact microphone*

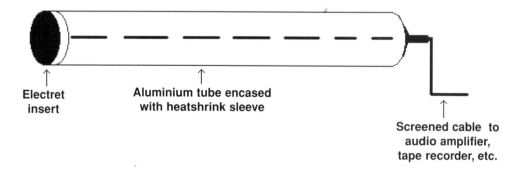

Figure 2.12   *Probe-type microphone*

and service entry/exit points, or a suitable hole may be formed by drilling.

Pen microphone construction

A suitable pen can be found in any large stationery supplies store. The pen can be a fat 'marker' pen or large-diameter pen, preferably black, with a flat-topped cap. The insides of the pen need to be removed, so choose one that is not all that messy to disassemble. The cap of the pen is simply drilled out so that it can accommodate the microphone, or the microphone may sit on the top of the pen as long as it blends in well, as shown in Figure 2.13. The microphone is

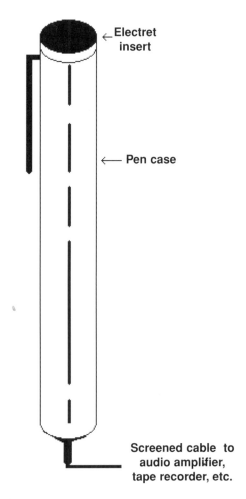

Figure 2.13   *Pen microphone*

soldered to a thin diameter length of screened cable which passes through the pen body and out through the 'nib' end. The microphone can be painted to match the rest of the pen casing once they have been glued together using superglue or fast drying epoxy resin adhesive.

The pen microphone really needs to be worn in the breast pocket of a jacket, to prevent muffled recordings or clothes rustling against it. This also means that a hole has to be cut in the internal layer of the pocket and jacket lining to enable the hidden cable (and jack plug if you were too eager to finish assembling the unit) to pass through to the recorder/transmitter device. It is always a good policy to form a knot or coil in the cable on the inside of the jacket, then sew this very firmly in place, just in case the target is someone who grabs a pen from the breast pocket rather than asking first. It would leave the operative very red-faced if the cable were suddenly to make an appearance. It may be a good idea to have one or two more identical dummy pens alongside the microphone one in case someone requests your pen for some reason.

If a pen microphone is not preferred, then a lapel or other body-worn microphone can be constructed by hiding a miniature microphone insert within the design of a brooch or badge. The cable is then passed through the clothing, so causing permanent damage to the article of clothing, unless a button hole can be used.

Audio frequency amplifiers

There is always room for high gain amplifiers in electronic surveillance. The output from a standard microphone can be poor and is easily boosted by amplifiers to make the microphone seem very sensitive. Units are available that are advertised as an aid for the hard of hearing. These units consist of a box looking rather like a personal stereo, which contains a sensitive microphone and high-gain audio amplifier circuit. The unit drives a pair of standard headphones or earphones. When used, the operative can place themselves near to the target, e.g. on an adjacent seat on a train, and listen in to conversations, if necessary, taping them. An amplifier can be used in conjunction with any discreet microphone system, whether it be in a complex CCTV with audio unit, or a pen/tieclip microphone.

Audio frequency amplifier designs

A selection of simple audio frequency amplifiers is shown on the next few pages. These amplifiers are suitable for many applications, but input and output impedance matching must be observed.

Simple audio frequency preamplifier

The first design, shown in Figure 2.14, which is for an audio preamplifier, uses only one transistor, and has an input that is suitable for electret microphones. If a dynamic or piezoelectric microphone is used, then resistor R1 is omitted. The component values need not be strictly adhered to, and the unit can be housed in a small plastic box, with a screened lead being used for the input connection. If space is at a premium, then the whole of the circuit, with the exception of the supply, can be housed inside a ¼" jack plug. The circuit has an output power of only approximately 50 mW, so cannot drive a speaker or a pair of headphones. The preamplifier does an excellent job when used between one of the electret microphone designs shown elsewhere in this book, and a tape recorder, to pick up all noises that may otherwise be missed by only using a microphone by itself. The supply voltage

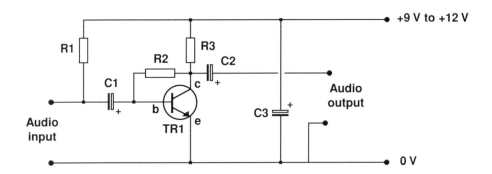

Component listing for one transistor audio amplifier

R1 = see text
R2 = 1M
R3 = 4k7

C1, 2 = 10 µF/18 V
C3 = 100 µF/16 V

TR1 = BC547

Figure 2.14   *One transistor audio amplifier*

requirement for this design is not critical and can be anywhere between 6 V and 15 V.

Audio preamplifiers using operational amplifier integrated circuits

Good quality, high gain preamplifiers can be built around the ever-present 741 op-amp IC. A d.c. blocking/audio frequency coupling capacitor should be used to connect from pin 6 of the IC to the following power amplifier/high impedance earphones if the former does not have one at the input. A design for a preamplifier that has a suitable input for piezoelectric devices is shown in Figure 2.15, and a design which has an input that is suitable for magnetic devices is shown in Figure 2.16.

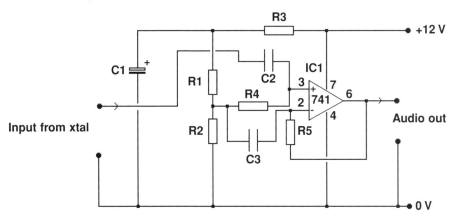

Component listing for crystal preamp

R1 = 18k   IC1 = 741 op-amp
R2 = 22k   C1 = 100 µf
R3 = 3k3   C2, 3 = 470 nF
R4, 5 = 100k

Figure 2.15 *Audio preamplifier for piezoelectric crystal input*

Low power audio amplifier

A low component count audio amplifier, capable of driving a pair of earphones or a small speaker, is shown in Figure 2.17. The circuit uses the popular audio amplifier IC, the LM386, which is supplied in an eight-pin package. The circuit works quite well with a supply voltage of 9 V, therefore the small, flat 9 V battery package can be used for small and lightweight portable/hand-held operations.

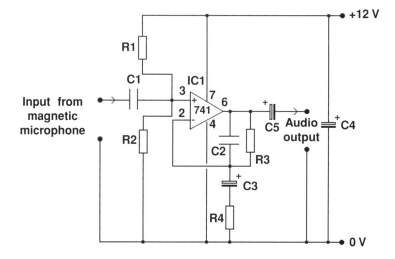

Component listing for magnetic microphone preamp

R1, 2 = 100k   C1 = 470 nF
R3 = 56k   C2 = 100 pF
R4 = 1k   C3 = 10 µF
                    C4, 5 = 100 µF

Figure 2.16   *Audio preamplifier for magnetic microphone*

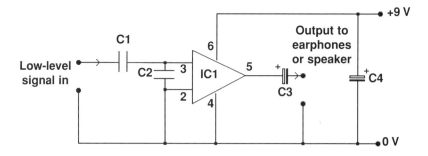

Component listing for low power audio amplifier

C1 = 470 nF
C2 = 270 pF
C3, 4 = 470 µF

IC1 = LM386

Figure 2.17   *Minimum component count low power audio amplifier*

Induction amplifier

An amplifier that can directly drive a pair of high impedance earphones, and that uses a coil as a pickup device, is shown in Figure 2.18. Resistor R4 acts as a gain control by altering the amount of feedback to the inverting input of the op-amp. The input device to the circuit can be a telephone pickup coil that is widely available from electronic suppliers, and intended to be stuck to the telephone set by means of a rubber sucker. Other coils, as well as other input devices such as piezoelectric contact microphones, can be experimented with. A coil from an old discarded electromechanical relay, for example, may have properties that allow a telephone call to be picked up by placing the coil against a telephone line. If the design were to be connected to a transmitter, then this would form the basis of a telephone tap that did not require any direct connection to the telephone line to be made, thereby making the task of planting a monitoring device very easy.

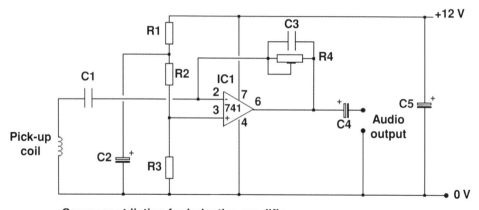

Component listing for inductive amplifier

R1 = 1k8	C1 = 100 nF	IC1 = 741 op-amp
R2 = 8k2	C2 = 100 µF	Coil = see text
R3 = 10k	C3 = 220 pF	
R4 = 10M pot	C4, 5 = 100 µF	

Figure 2.18   *Inductive tapping amplifier*

Already got an audio amplifier?

The reader may already have a ready-built useful amplifier in their possession. Some of the older 'piano-key' style cassette tape recorders have the quite often undocumented feature of listen-through. Try plugging a pair of headphones into the earphone socket, then pressing the record button. The built-in microphone may pick up noises that are amplified to the headphones. On some models, this listen-through may only work whilst the record and pause buttons are both pressed down. With the addition of an external microphone, with or without a preamplifier in line, this quick and easy method may prove effective. Other newer styles of recorder may have this feature, which can be tried out in an electronic store before purchase.

Conversation amplifier

A simple three-transistor conversation amplifier circuit is shown in Figure 2.19. The unit can be built so that the input microphone, in this case an ordinary speaker, can be held near to the target conversation, or hidden in a room. Another conversation amplifier circuit design, this time using an electret microphone and only two transistors, is shown in Figure 2.20. The microphone may be a probe microphone,

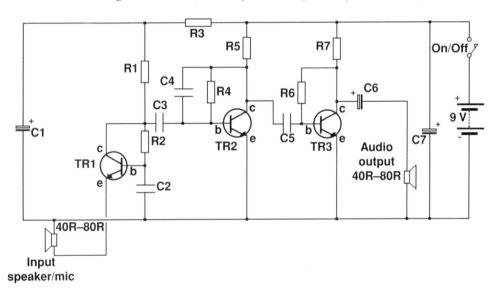

Component listing for conversation amplifier

R1, 5 = 4k7	C1, 6, 7 = 100 µF
R2 = 680k	C2, 3, 5 = 220 nF
R3, 7 = 470R	C4 = 330 pF
R4 = 1M5	
R6 = 47k	TR1, 2, 3 = BC547

Figure 2.19   *Conversation amplifier*

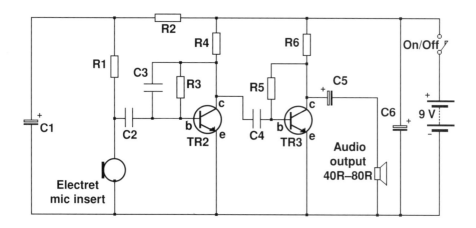

Component listing **for conversation amplifier with electret microphone**

R1, 4 = 4k7	C1, 5, 6 = 100 μF
R2, 6 = 470R	C2, 4 = 220 nF
R3 = 1M5	C3 = 330 pF
R5 = 47k	Tr2, 3 = BC547

Figure 2.20 *Conversation amplifier with electret microphone*

built into a parabolic dish, for example. In both cases, a volume/sensitivity control can be added by placing a suitable variable resistor from the speaker end of the output capacitor to ground, with the output picked off to the speaker or earphones.

Cameras

CCD cameras have enabled video surveillance techniques, both fixed and portable, to go from strength to strength in the last few years. With small diameter (3 mm) wide angle lenses, with a small circuit board that can produce a standard 1 V peak-to-peak video output, it means that the lens, infrared light emitting diodes (for seeing in the dark), and board can be put into a housing of around 35 mm square. Power requirements may be satisfied by a standard PP3 9 V battery.

Cameras hidden inside fully functioning smoke detector alarms, badges, books, buckles, brooches, etc. can be produced. The camera wiring required for fixed and portable use is minimal, with just a screened two-core wire supplying operating power from the battery, and the standard video signal from the camera going to a concealed miniature video recorder. The video signal obtained from a camera can easily be connected to the input of a miniature video transmitter, also battery powered if required, to make a completely portable surveillance unit about the same size as a pack of 20 cigarettes. Light-bending pipes (fibre-optic tubes) can be used in conjunction with lenses to obtain video information from awkward angles, holes drilled in walls, etc.

Carrier current transmitters

For several years now, this device has been successfully utilized in baby alarms and 'wire-less' intercommunication systems, and have found their way into surveillance equipment design. These baby alarm devices are comprised of a transmitter unit and receiver unit. Both units are plugged into the mains supply sockets and instead of radiating radio power through the ether, they use the mains supply wiring as a transmission medium. Typically these units produce an RF signal, frequency modulated in the region of 150–300 kHz. This modulated radio frequency signal is usually capacitor coupled to the mains wiring, with a range of a few metres or so, enough to cover an average sized building.

In theory, if enough radio frequency power is coupled to the mains wiring, then the information would be transmitted a much further distance, up to any power company transformer. Some supply companies include RF bypass capacitors across their transformers because commercial use of carrier current transmitters is often implemented for data transfer and switching purposes.

A surveillance device using this technique can be constructed to give a transmission that is sufficiently powerful to be received by the operative, but not so strong as to be detected at a greater distance. The signal produced from some commercially available baby alarms can be picked up from a very short range if a long wave radio is in the vicinity, especially if the receiver is placed in close proximity to the buildings' mains supply wiring. The clandestine carrier current transmitter device will operate indefinitely since it is mains powered and can be built and concealed in places similar to those of a standard mains powered radio frequency transmitter. Due to the fact that they use the mains supply cable as a transmission medium, these devices should be relatively easy to detect and pinpoint.

Modulation

Up to now, all transmitting devices mentioned have been spoken of as modulated by voices or noises. The standard method of modulation is that of 'frequency modulation', or FM. As a method of modulation has distinct advantages, in surveillance devices, since compared to 'amplitude modulation', the following apply:

1 Far less power is required to frequency modulate a transmission to give acceptable recovered audio at the receiver. This generally means a lower battery drain.
2 A lower component count is required, enabling the transmitting device to be built much smaller.
3 It is generally accepted that FM reception is more cleaner than a more noise-prone AM (amplitude modulation) system. Given that the transmission from a surveillance transmitter is very weak, it is imperative that any signal received by the operative is not swamped by electrical interference.

In frequency modulation, a fixed carrier frequency, of say 100 MHz (one hundred million cycles per second), is modulated by the information from the audio circuit. With a commercial VHF FM music radio station, the amount of deviation from the fixed carrier frequency of 100 MHz would be in the region of 75 kHz. Since high fidelity reproduction is no concern of the designer of a surveillance transmitter, the amount of deviation required to convey acceptable speech is kept to a minimum, therefore a deviation level of 5–10 kHz will more than suffice for this purpose. The bandwidth of many telephone circuits is only a few kilohertz, as this is thought sufficiently wide enough to be capable of conveying intelligible speech, therefore any surveillance device designed for conveying telephone conversations does not require excessive deviation.

Transmission frequencies of surveillance transmitters

There are no rules to the frequency that surveillance transmitters operate on, but general guidelines are as follows.

Many budget 'disposable' transmitters are tuneable and operate between 88 MHz to 108 MHz or up to 135 MHz. The reason for this is that a standard commercially available VHF/FM receiver will cover the lower range, with a standard 'airband' receiver covering the higher frequency range. The advantages and disadvantages of using the standard commercial band (88–108 MHz) for covert surveillance transmitters are:

1 Receiving equipment, complete with tape recorders if required, are always cheap and readily available.
2 With receivers having very good selectivity, it is possible to tune the output of a surveillance transmitter right alongside a powerful commercial channel so the weak surveillance transmission can hide under the skirts of the commercial channel. This means that any casual scan through the band will skip from one commercial channel to another, riding passed the relatively weak signal.
3 Unless transmissions are weak, or 'hidden' as previously described, due to the many people who have receivers capable of tuning through the domestic band, accidental discovery is imminent.
4 Many common portable receivers have poor sensitivity, since they are designed and built only to receive signals that are relatively strong, to be received by in-car units. The output from a standard budget surveillance transmitter will in comparison be extremely weak, both in power and deviation level.

The advantages and disadvantages for using a system on the 'airband' are:

1 Simple receivers meant for the reception of the airband frequencies are generally of poor sensitivity.
2 The range of dedicated airband receivers are designed to receive AM signals. Surveillance transmitters are designed to produce an FM signal. Although some of the simpler transmitter designs give out a signal that is a mixture of FM and AM, reception on a dedicated AM airband receiver may be of poor quality. The method of 'slope detection' then has to be employed.
3 Low power transmitting devices should only be used, since a transmission range of 500 m in a built-up urban area may be quadrupled when there are no obstacles between the radiating aerial and a low-flying aeroplane. This situation could have serious consequences!
4 As mentioned above, the airband is used not only for air/ground and ground/air communications, but is also shared by other commercial companies, etc. Any kind of interference, be it from a legal or illegal source, is considered a gross offence.
5 There are a very large number of airband listening enthusiasts, several of whom use high-gain aerial systems in pursuit of their hobby. The chances of accidental discovery of a 'covert' signal is therefore possible.

6 In any geographical area, there is always an abundance of frequencies that remain permanently quiet, making an ideal spot for monitoring uninterrupted transmissions from surveillance transmitters.

Many other frequencies are used for transmission. These frequencies are used for 'professional' devices, which are not tuneable but use crystal controlled oscillators, with one or more switchable channels. The operational frequency ceiling is around 450 MHz. The reason for this ceiling is that frequencies above this are used by security organizations, then above these channels, UHF terrestrial television channels appear. Any device operating near a television channel could well produce patterning and other types of interference to television reception, as well as interference to the operative trying to receive data from a surveillance unit.

Some devices operate at frequencies well above 1 GHz. The disadvantage of using higher frequencies is that the 'penetrating power', i.e. the ability of the radio waves to travel through the fabric of buildings and other obstacles, is greatly reduced.

To summarize, the frequencies most often used by common surveillance transmitters are:

- 88–108 MHz tuneable VHF 'amateur' low-cost disposable types.
- 108–140 MHz tuneable or crystal controlled VHF 'airband' types.
- 365–455 MHz 'professional' crystal controlled UHF.

It must be emphasized that any frequency can be used by transmitting devices. Surveillance devices remain covert because trends are always changing.

Sub-carrier transmitters

The method of modulation mentioned above, frequency modulation, has a variation, which may be described as frequency modulated frequency modulation, or FM/FM. In this system, sometimes referred to as a 'sub-audio', 'ghost' or 'phantom' device, modulation is carried out as follows. Instead of modulating the radio frequency carrier, e.g. 100 MHz, in a direct fashion, the audio signal, e.g. 5 kHz, is injected into a sub-carrier oscillator, operating on a frequency of say, 60 kHz. This 60 kHz signal then modulates the 100 MHz carrier frequency. Anyone now tuning to 100 MHz on a standard FM receiver would detect the carrier but would only hear a hissing through the loudspeaker. To recover the modulating audio from such a transmission requires the use of a decoder or demodulator unit integrated into the receiver.

Since a wide range of sub-carrier frequencies, for example, anything from 25 kHz to 90 kHz, may be adopted, and since a dedicated network of filters is required for any specified sub-carrier frequency, this system is relatively secure unless expensive receiving equipment is at hand. Note that although the audio information of sub-carrier transmissions are not easily stumbled upon 'accident-ally', the radio transmission itself can be picked up by standard RF detection methods.

In the design of a sub-carrier transmitter, it is a requirement that there is no modulation of the carrier frequency whatsoever by the original audio information signal, since even the slightest leak of audio appearing on a transmission could be resolved.

Due to their large component count and the requirements of a dedicated decoder circuit, such transmitting devices are relatively large, and since the system package includes a receiver with decoder, the units are more expensive than their 'direct frequency modulation' counterparts.

Optical communication

In recent years, the availability of low-cost, high-powered invisible light (infrared) devices has meant that these units are now being used not only to provide endless hours of enjoyment from the use of cordless headphones (where a transmitter is plugged into the head-phone jack of a television or record player, and the IR receiving unit is attached to the headphones), but also in the field of surveillance. Although somewhat limited in range, and prone to interference from any other infrared source such as fires, etc. some devices can utilize this method for communicating information. An infrared beacon may be surreptitiously placed on a vehicle, boat, etc. and with the aid of an infrared camera or binoculars will enable the target to be spotted and followed at night with great ease.

The theory of lasers used as voice interceptors

Laser technology has been used for eavesdropping purposes since the last quarter of the twentieth century. The method used is briefly outlined in Figure 2.21. A laser tube is used to project a beam of light for a required distance after it has passed through a network of lenses so as to correct the beam, ensuring that it is perfectly parallel and that

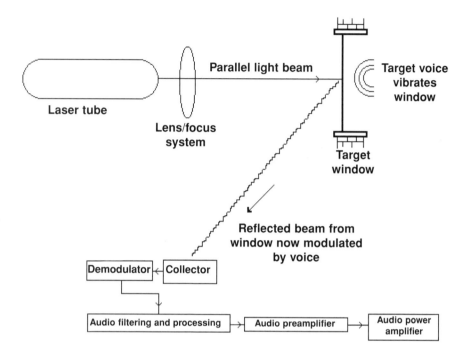

Figure 2.21 *Theory of lasers as voice interceptors*

it is as narrow as possible. By the use of a gunsight or other aiming device, the unmodulated laser beam is directed at the target window, on the other side of which a conversation is hopefully taking place. The air pressure inside the room will alter in sympathy with the sound, therefore the window will also vibrate. This means that the laser beam that is already being reflected back to the listening station will now be a carrier beam that is slightly modulated by the vibrations from the window. The received modulated beam is demodulated so that the modulating audio content can be recovered, very heavily filtered so that unwanted noise is removed, then the clean audio is then amplified so that it can be listened to by the operative.

Although visible and non-visible wavelengths for the laser device are available, both may suffer from external sources of interference. Even if positioned just a few yards away, from across the other side of the road, for example, smoke and dust particles from passing traffic will cause the transmitted beam to be attenuated, corrupted and diverged. If used at night, the visible laser may be spotted as it lights up smoke particles, etc. in the air, when it travels across a street, whereas a long-distance infrared system would be corrupted by sunlight in the daytime and certain artificial lighting at night. The amplitude of voice modulation of the received signal is dependent upon a wide range of factors, including how well the glazing of the target building is soundproofed. The threat of laser surveillance is taken very seriously by the security authorities, which was proved by a certain intelligence centre HQ in London, England, when it had to reinstall new triple glazed windows in its brand new building, causing an uproar amongst the taxpayers when tests were performed and it was then found out that the double glazed units were of insufficient sound proofing against 'enemy' laser eavesdropping technology.

Noise filters

In days gone by, noise filtering was a state-of-the-art technology in the field of eavesdropping. The aim of filtering out noise from either live (hard to do) or recorded (easier to do) conversations was to filter out any unwanted noise that was making the intercepted conversation either unintelligible so that it was not easy to catch the general drift of the conversation, or to clean up the recording so that it would make more acceptable admissible evidence in a court of law. Expensive audio filters were needed to clip out deep rumbles of heavy street traffic, or the high pitched squeak of a machine, as shown in Figure 2.22.

Nowadays, computers are used for the task of filtering out unwanted noises, taking just a matter of a few minutes instead of several days, notching out rubbish with a high degree of accuracy and completion, compared to the old-fashioned hardware method that used to take days to complete without the same measure of success. Software packages originally intended to work on cleaning up musical recordings on scratched vinyl, or for 'sampling' music sections, are widely used for this new task with a great deal of success. The example shown in the illustration has points in the waveform marked 'X' which denote noises that are detracting from the recorded conversation, but with the aid of suitable computer software, they have been highlighted and selected for zero amplitude, thereby removing them from the surveillance tape recording. To save having to record the conversation on tape, then dump it into the computer, it is now possible to compress the size of the audio information from the source and use the computer to record the information directly. By cutting out the 'middle-man', the quality of recording can be much better.

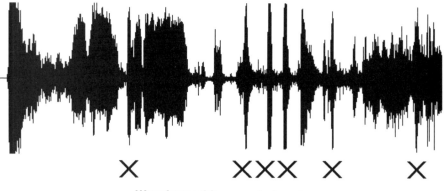

Waveform with unwanted peaks

Waveform with unwanted peaks cut out by PC software

Figure 2.22 *Noise filtering – PC controlled example*

Retransmission of an audio signal – relay transmitters

As mentioned previously, it is sometimes necessary to find some way of boosting the transmitted signal that is generated by a clandestine device. This may be required if the transmission from, say, a weak telephone transmitter or low powered room transmitter, cannot be locally monitored by the operative. This may be due to the possibility of being caught out whilst being in the close proximity of a sensitive area, or by the fact that the monitoring station is quite some distance away from the target area. If this is the case, then it is necessary to design a system where the weak, local signal is picked up by a nearby receiver, then retransmitted by a more powerful relay transmitter, as illustrated in the block diagram in Figure 2.23. The relay station may be situated either inside the building in an area that is not as sensitive as that where the low powered device is situated, such as a garage, storeroom, etc. or in a parked vehicle outside the building or complex.

The transmission from the hidden transmitter must be picked up by a good receiver that is capable of self-adjusting to any frequency drifting of the transmitter, since the receiving unit will be remote and frequency tuning would therefore not be an option open to the operative. The choice of a relay transmitter is somewhat critical, with regards to power level and frequency of transmission, although any frequency drifting would not be so much of a problem because the remote monitoring centre would be able to adjust their receiver to allow good copy. It should be remembered that if a high powered relay transmitter is being used, then there is a good chance of interception.

An alternative to using a high powered relay radio transmitter is to plug the audio of the monitoring receiver into the microphone input of a cellular telephone, as shown in Figure 2.24. In some countries, the transmission system for cellular telephones is a digital one, which in theory is not possible to intercept by a third party. Practical

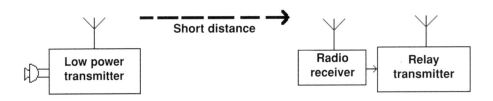

Figure 2.23 *Retransmission of a signal using a relay transmitter*

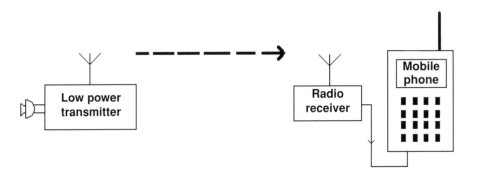

Figure 2.24 *Retransmission of a signal using a cellular telephone*

considerations for using this method are that the telephone handset needs to have either a mains power source to keep the battery topped up, as does the receiver, or a large capacity lead–acid battery, such as the type used in automobiles, to provide a long service if so required. Another problem with using a cellular telephone is that the (large) bill has to go to someone, and if the system is ever discovered, the bill may be traced back by the security services to the operative, or at least, the owner of the telephone number.

A typical interface device that may be used to connect the earphone output of the receiver to the microphone input of a relay transmitter or telephone is shown in Figure 2.25. The problems that may be encountered are RF from the high powered relay transmitter entering the microphone input causing feedback, and impedance matching between the receiver output and transmitter input. All leads should be made from screened cable, and the unit should be built into a metal enclosure. To minimize RF, decoupling capacitors C1 and C2 are employed. Audio coupling/d.c. blocking capacitors C3 and C4 allow the audio a pathway, but provide d.c. isolation between receiver output and transmitter input, whereas diodes D1 and D2 act as clipping diodes to prevent large voltages, which may be generated in RF fields, from damaging components in the microphone input circuitry of the transmitting device. R2 acts as a volume control, so as to find a correct audio level to the transmitter audio input.

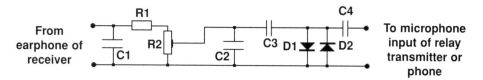

Component listing for radio receiver to relay transmitter interface

C1, 2 = 1 nF
C3, 4 = 100 nF
D1, 2 = 1N4007
R1, 2 = 1M

Figure 2.25 *Typical interface for receiver to relay transmitter or cellular telephone*

Tracking beacons

Tracking beacons are used to find the whereabouts of a portable or mobile object, for example a motor vehicle, stolen equipment and plant, wild animals, ferrets used for rabbit hunting, pet dogs, etc. upon or inside which the transmitter has been planted. There must be some way to monitor the transmissions emitted, and the system requirements are illustrated in Figure 2.26. A directional aerial is employed, which can be rotated so as to find the direction of the transmitter. Antennae properties vary greatly depending upon their construction.

Directional aerial system

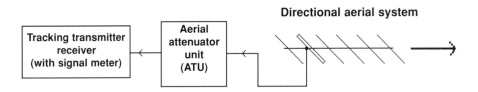

Figure 2.26   *System requirements for tracking transmissions*

The simplified polar patterns or diagrams of an assortment of aerials are shown in Figure 2.27. The first example is a basic dipole aerial, and by looking at the polar pattern it is possible to see the drawback of using this simple aerial for direction finding, i.e. both the forward and reverse gain are identical, in fact the polar pattern for a vertically polarized dipole, if drawn in three dimensions, would represent a doughnut shape. The simplicity of the dipole, however, should not be ignored, since if, for example, the target vehicle was within a certain radius of an event, and that the direction that the target vehicle approaching was not known, then a non-directional, or omnidirectional dipole would be employed. The next aerial shown is a dipole with a single added element. This is a slight improvement over a simple dipole since the forward and reverse gain are slightly different, which means that if time is allowed to rotate the aerial fully 360 degrees, it will be noted that the direction can be deduced. The third example shown is a dipole with one reflector element and four director elements attached. This aerial would give a much better idea of the direction of the transmitter, since the reverse gain is far smaller than the front gain, although as an aerial of this nature increases in complexity, little side nodes or lobes will appear on the polar diagram. This means that the aerial has to be worked by an experienced operative, rotated fully back and forth until they are fully confident that the main frontal lobe is pointing at the target and that a weak signal is not in fact the target twenty feet away from them in a weak rear lobe.

Although a good directional aerial with a sharp main front lobe can find the direction of a tracking transmitter, or a hidden room transmitter, etc. and an experienced operative will have a relatively

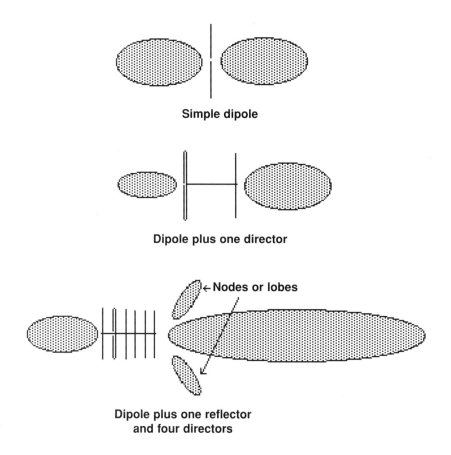

Simple dipole

Dipole plus one director

←**Nodes or lobes**

**Dipole plus one reflector
and four directors**

Figure 2.27   *Simplified polar patterns for various antennae*

good, educated guess at the range between the receiving aerial and transmitter, a problem will occur when the transmitter gets close to the receiver. This is because since no matter in which direction the aerial is turned, the transmitting signal will overload the signal strength meter on the receiver. To avoid this overload problem, an attenuator is connected via screened cable between the receiving aerial and receiver. The attenuator can be switched in steps, but preferably would be continuously variable in nature. Whenever the distance between the transmitter and receiver is too close, where the signal strength meter is about to go off the scale, the attenuation between the aerial and receiver is increased so that the signal reading is more acceptable.

Another method of tracking and direction finding is to use a nest of quarter wave aerials on top of a vehicle. With the outputs from the aerial array connected to a computer-controlled receiver, the microscopic time difference it takes for the same signal to be picked up by the aerials only a metre or two away from each other is measured, so by computation, the direction of the transmission can be readily established with pinpoint accuracy almost in an instant.

A simple tracking transmitter that uses just one cell for power is shown in Figure 2.28. The circuit can be built very small and was originally designed for tracking ferrets during rabbit hunting at night, hence the LED which will emit a bright flash every second or so. The output from the flasher IC, the 3909, is also connected to a single

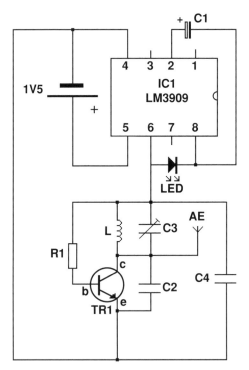

Component listing for 1V5 tracking beacon

R1 = 10k	IC1 = 3909 flasher IC
C1 = 100 μF	TR1 = BC547
C2 = 4p7	L = 6 turns of 22 swg enamelled copper wire
C3 = 2–10 pF	wound on ¼″ former
C4 = 1 nF	LED = LED to suit

Figure 2.28   *Single-cell 1V5 interrupted carrier-only beacon transmitter*

transistor VHF oscillator. A 1 nF decoupling capacitor may have to be connected between the base of TR1 and ground if the unit fails to oscillate. The output of the transmitter is carrier-only, so a receiver with a BFO is required.

Another tracking transmitter is shown in Figure 2.29. The 555 timer will give a constant audio output signal that is connected to the single transistor transmitter, producing a steady tone from the loudspeaker of the radio receiver. Modulation derived from the audio oscillator is more than ample, therefore the modulation level is controlled by the carbon pot R3. If a tracking transmitter is required to give out pulses of tone, as seen on television, then this is performed by the circuit design shown in Figure 2.30. The audio tone oscillator IC, that is IC1, is controlled by the output of another pulsing circuit that is high for one second and low for 200 ms, and is performed by IC2 and associated components. Neither of these last two tracking transmitter requires a receiver fitted with a BFO because they both transmit a frequency modulated signal that can be picked up on a standard VHF FM radio.

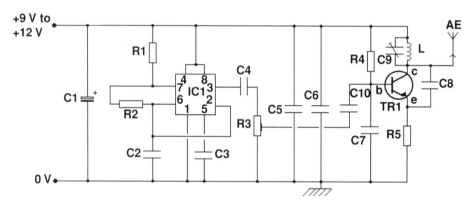

Component listing for automobile tracking transmitter

R1 = 4k7	C1 = 100 μF	IC1 = NE555 timer IC
R2 = 1k	C2, 3, 4, 6, 10 = 100 nF	TR1 = BC547
R3 = 1k pot	C5, 7 = 1 nF	
R4 = 100k	C8 = 4p7	
R5 = 560R	C9 = 4–22 pF	

L = 6 turns of 22 swg enamelled copper wire wound on ¼" former
AE = aerial constructed from approx. 12" of insulated hook-up wire

Figure 2.29 *Automobile 'tracking' transmitter*

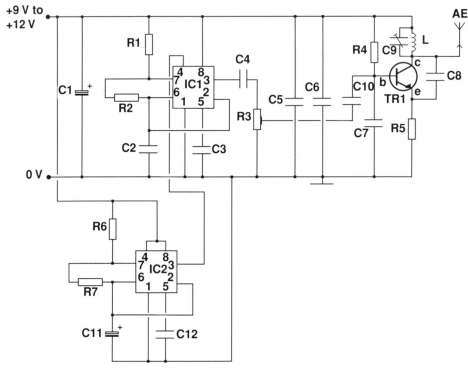

Component listing for automobile tracking transmitter with bleeping audio

R1, 2, 7 = 10k
R3 = 1k pot
R4, 6 = 100k
R5 = 560R
C1 = 100 μF
C2, 3, 4, 6, 10, 12 = 100 nF
C5, 7 = 1 nF
C8 = 4p7
C9 = 4–22 pF
C11 = 10 μF

IC1, 2 = 555 timer IC
TR1 = BC547
L = 6 turns of 22 swg enamelled copper
    wire wound on ¼″ former
AE = aerial constructed from approx.
    12″ of insulated hook-up wire

Figure 2.30 *Automobile 'tracking' transmitter with bleeping audio*

3 Room transmitters

Before considering the different designs of room transmitters, it should be noted that the basic building block of any transmitter is the oscillator. The oscillator section of a transmitter is the circuit portion that defines the working frequency of the device.

In surveillance devices, the oscillator will be one of two types:

1 the 'VFO', variable frequency oscillator, or 'free-running' oscillator;
2 the crystal controlled oscillator.

Both types of oscillator have good and bad points.

The VFO

The VFO oscillator will operate at the output frequency of a device, i.e. a unit whose output frequency is 110 MHz will have the VFO also operating at 110 MHz. This is because the VFO, unless constructed to the very highest standard, is inherently unstable. If a large object comes near to the unit or unit aerial, there is every chance that this will cause the VFO to go off-frequency by varying degrees. The frequency output of a VFO will also be altered whenever the oscillator is loaded, i.e. if excessive power is drawn out of it due to a serious impedance mismatch. Also, the VFO is prone to be affected by temperature fluctuations. Due to the fact that the oscillator generally operates at the transmission frequency, the amount of deviation/ frequency modulation is generally low, since the amount of frequency deviation is not multiplied by any following stage of buffer or radio frequency power amplification.

The basic electronic components that control the operational frequency of a VFO are a coil (L) and capacitor (C), which are connected to form an 'L/C' combination. To alter the frequency of the combination, either the coil or capacitor are variable. If a coil is to be variable, it can be a coil of copper wire that can be compressed or stretched to alter the electrical characteristics, or preferably be of the iron dust core type that is screwed in or out of the coil former. This requires the use of a proper plastic trimmer tool to avoid the core disintegrating. The very act of putting a hand near to a VFO coil will send the output frequency haywire. If a variable capacitor is used to

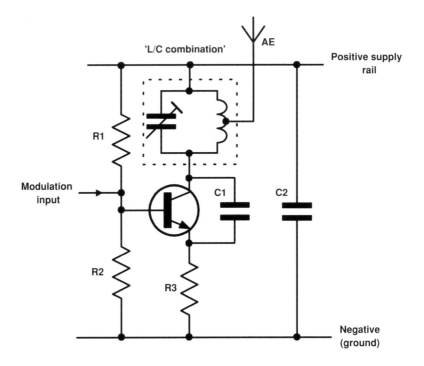

Figure 3.1   *Basic radio frequency oscillator*

adjust the frequency, miniature types are used, rather than the very bulky 'postage-stamp' design.

The advantages of the VFO are that as the transmitter usually operates at the frequency of the oscillator, no complex frequency multiplication circuits are required, leading to relatively low cost and small construction. Since no multiplication stages are used, this may make accidental discovery less possible, since no frequencies below that of the basic oscillator are produced by the transmitter. Due to the 'single stage' design, power consumption (current drain) from a battery supply can be very low, sometimes in the order of only a few milliamps.

Figure 3.1 shows a typical simple VFO used in a basic surveillance transmitter. Base bias is produced from R1/R2, and RF decoupling across the supply lines is provided by C2. The parallel tuned circuit, the 'L/C combination', is designed to produce the required output frequency of the transmitter. To maintain oscillation, feedback is supplied by C1, which is connected between the collector and emitter of the transistor. Modulation to the oscillator is fed to the base of the transistor, through a capacitor either with or without a series resistor.

The crystal controlled oscillator

The crystal controlled oscillator has the ability to produce a signal on one particular frequency, with an accuracy of a few parts per million. Although it is possible to obtain a crystal that can operate on high frequencies in transmitters, it is more usual to construct an oscillator working at a base frequency, e.g. 4 MHz, followed by multiplier stages to produce the final frequency of the transmitter. Modulation is applied directly to the oscillator, in the same fashion as the VFO, but since the frequency generated by the crystal oscillator is then multiplied, so is the amount of frequency deviation. For example, if a 4 MHz oscillator is pulled by the modulation by 500 Hz, and there is a multiplication factor of:

$$4\,\text{MHz} \times 3 = 12\,\text{MHz}$$
$$12\,\text{MHz} \times 3 = 36\,\text{MHz}$$
$$36\,\text{MHz} \times 3 = 108\,\text{MHz}$$

then the deviation would be:

$$500\,\text{Hz} \times 3 = 1\,500\,\text{Hz}$$
$$1\,500\,\text{Hz} \times 3 = 4\,500\,\text{Hz}$$
$$4\,500\,\text{Hz} \times 3 = 13\,500\,\text{Hz, or } 13.5\,\text{KHz}$$

This amount of deviation would produce almost 'high-fidelity' audio from a covert transmission.

This creates a transmitter with stable frequency output and with more than adequate deviation for the transmission of room conversations. There is a price to pay for quality. The crystal controlled oscillator requires more components than a simple VFO. Each stage must be tuned carefully to avoid tuning to an incorrect frequency produced by any preceding stages. The cost and physical size of a crystal controlled device will be larger than a VFO design, as will the current consumption from the power supply. Since frequencies other than the output frequency are also produced, as seen in the example above, radiation on these frequencies can be detected unless each stage is physically screened, and an high pass filter is connected to the aerial output socket of the transmitter.

An example of a crystal controlled oscillator and associated multiplier stages is shown in Figure 3.2. TR1 and associated circuitry form the crystal controlled oscillator. D.c. biasing is derived from R1, R2 and R4. Feedback is obtained via the 'RF tap' between C1 and C2. The oscillator crystal is a 12 MHz device, and output from the oscillator is fed to the following multiplier stages of TR2 and TR3. The

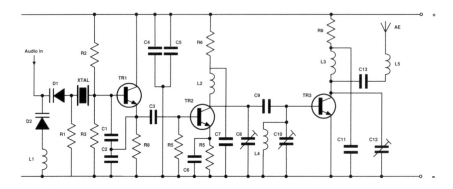

Figure 3.2   *Crystal controlled oscillator and multiplier stages*

12 MHz signal is first fed into a frequency tripler to produce 36 MHz, then into a further tripler to yield 108 MHz. Modulation from a microphone amplifier circuit is fed to the varactor (varicap) diode input. This special diode has the ability to swing its built-in capacitance whenever a reverse bias voltage is placed across it. Since any change in capacitance across a crystal will alter the operating frequency slightly, the varying capacitance of the varicap diode alters the output frequency of the circuit, i.e. it is then frequency modulated.

Component listing for crystal controlled oscillator and multiplier stages

Resistors

R1	68 k	R5	120 R
R2	15 k	R6	68 R
R3	15 k	R7	4 k7
R4	1 k5	R8	120 R

Capacitors

C1	100 pF	C7	10 nF
C2	100 pF	C8	5–100 pF
C3	470 pF	C9	10 pF
C4	10 nF	C10	5–100 pF
C5	22 μF	C11	10 nF
C6	10 nF	C12	2–22 pF
		C13	5 pF

Inductors

L1	4.7 μH	L3	0.1 μH
L2	0.22 μH	L4	0.22 μH
		L5	0.1 μH

Semiconductors
TR1, 2, 3   ZTX327
D1, 2   BA102

Quartz Crystal
12 MHz for 108 MHz output

Simple voice transmitter

The circuit diagram of a basic simple voice transmitter can be seen in Figure 3.3. The simplicity, cheapness and ease of use has made this and similar circuits extremely popular for several years. The circuit is almost identical to that shown in Figure 3.1. An electret microphone insert replaces base biasing resistor R2. Since the audio signal modulates the steady RF carrier frequency, a modulated transmission results. The modulation from this circuit is probably rather a mixture

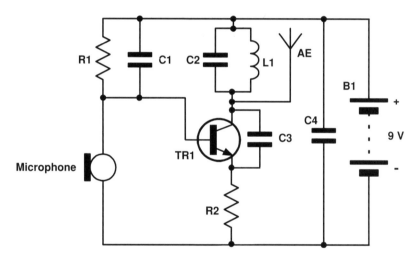

Figure 3.3   *Simple voice transmitter*

of frequency modulation with a bit of amplitude modulation thrown in for good measure, but despite this, the signal is easily readable on a VHF-FM receiver.

Component listing for simple voice transmitter, Figure 3.3
Resistors
R1   15 k   R2   330 R

Capacitors

C1	1 nF	C3	5 p6
C2	5 p6	C4	1 nF

Semiconductors

TR1	BC547

Microphone
Electret microphone insert, 10 mm diameter
2-terminal type

Inductor
5–6 turns of 22 swg enamelled copper wire
6 mm diameter, air spaced

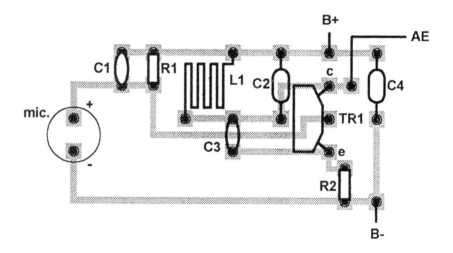

Figure 3.4a   *Simple voice transmitter component layout*

The circuit can be built on a printed circuit board as shown in Figures 3.4a and 3.4b, or on a small piece of stripboard. The aerial wire should be approximatley 60 cm of insulated wire. The unit in Figure 3.4, drawing around 25 mA of current from a 9 V PP3 style battery, can have a range of around 100–500 m. To tune the transmitter, it is necessary for the coil to be gently stretched or compressed. If required, the coil can be replaced with a four-turn manufactured coil with slug tuning, or the 5 p6 capacitor in parallel with the coil may be replaced with a 5 pF trimmer capacitor. Battery life can be extended

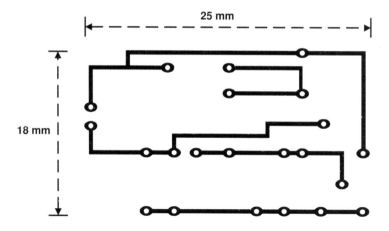

Figure 3.4b *Simple voice transmitter printed circuit board*

by increasing the emitter resistor, with a decreased transmission range resulting from this modification.

As previously mentioned, the problem found when modulating a VFO is that the amount of modulation/deviation is relatively low, since there are no multiplier stages following on after the oscillator section. This can be overcome to some extent by increasing the level of audio to modulate the VFO. Using E-line transistors and small support components, e.g. 0.125 W resistors, or 'SMD' techniques, the problem of keeping the unit small can be easily overcome. Figure 3.5 shows a transmitter using a microphone amplifier, where TR2 and associated components act as an audio amplifier, with coupling to the VFO being via R4 and C3.

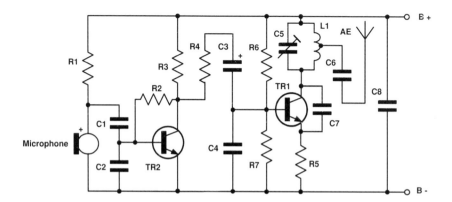

Figure 3.5 *Audio transmitter with microphone amplifier*

Component listing for transmitter with microphone amplifier, Figure 3.5

Resistors

R1	82 k	R3	15 k
R2	470 k	R4	4 k7
R5	470 R	R6	33 k
		R7	33 k

Capacitors

C1	22 nF	C5	2–10 pF
C2	1 nF	C6	1 nF
C3	10 μF	C7	5 p6
C4	1 nF	C8	1 nF

Semiconductors

TR1, 2   BC547, ZTX300, etc.

Other components are as per Figure 3.3, except that the aerial coil is tapped one turn from the junction with TR1 to provide an impedance match.

Note that since the value of R4 will control the amount of deviation that is superimposed upon the oscillator section, and that the amount of deviation, to some measure, depends on the gain of TR2, the value of R4 may need to be altered by experimentation. To maintain the correct amount of deviation, the value of R4 should be increased if the operational frequency of the transmitter is increased.

Mains powered supplies

Typical mains powered supplies can be seen in Figure 3.6 and Figure 3.7. In Figure 3.6, the circuit relies on the property of capacitors called

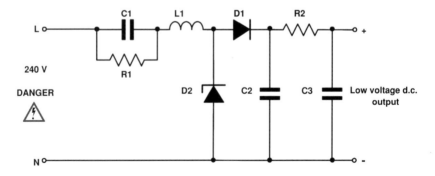

Figure 3.6   *Mains power supplies for miniature surveillance transmitters*

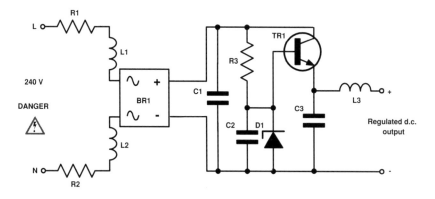

Figure 3.7 *Standard RF filtered d.c. supply for miniature surveillance transmitters*

reactance, where the capacitor acts as a 'variable resistor' whose resistance changes at different frequencies. The reactance of the capacitor, C1 (which is a high working voltage $0.1\,\mu F$), at 50 Hz, is approximately 32 000 ohms, i.e. it looks like a 32 kohm resistor to the 50 Hz mains frequency. Diodes D1 and D2 supply voltage regulation and rectification, with the 'T' formed by C2, R2 and C3, giving smoothing to provide clean d.c. to the transmitter circuit. Figure 3.7 shows a circuit, which apart from the voltage dropping resistors R1 and R2, is a standard RF filtered d.c. supply used in poor quality power supplies. Since a lot of effort has gone into this latter design to provide RF filtering, it is possible to couple the radio power from the final stage of a transmitter (see Figure 3.8), and by using a high

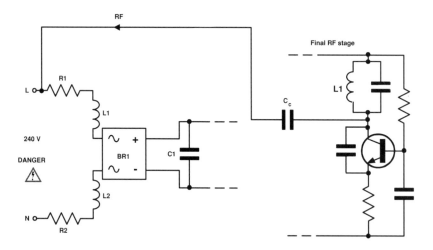

Figure 3.8 *Method of coupling RF energy to mains wiring*

voltage capacitor of a few picofarads, inject the RF into the mains supply and use the mains wiring as a super-long radiating element.

Simple mains powered voice transmitter built inside a double outlet adaptor

Figures 3.9, 3.10, 3.11 and 3.12 show a complete, simple mains powered transmitter that, with patience, can be built inside a functioning two-way mains adaptor. The transmitter section is a copy of the circuit shown in Figure 3.3. The extra components required are:

- C5, 6 $100\ \mu$F/16 V electrolytic
- C7 $0.1\ \mu$F/250 V poly
- R4   470 k
- R3   100 R
- D1   1N4148
- D2   5–8 V zener diode

Note that although an aerial is shown, enough RF power may trickle into the mains wiring to make this unnecessary. The unit is built in three sections, the transmitter board, C1, R1 and L1, then D1, D2, R2, C2 and C3. The only modification required to be carried out upon the adaptor is a small notch or hole that has to be made in the internal wall to pass the '+' (positive) and '–' (negative) leads from C3 to the miniature transmitter board.

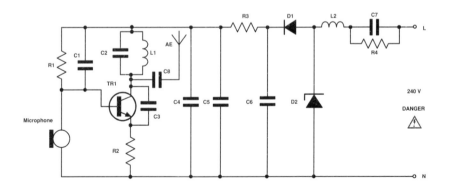

Figure 3.9   *Simple mains powered voice transmitter*

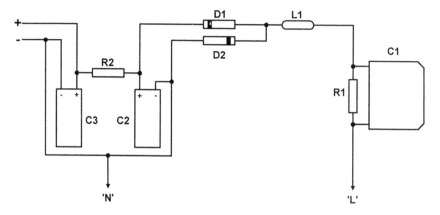

Figure 3.10 *Mains adaptor supply component details*

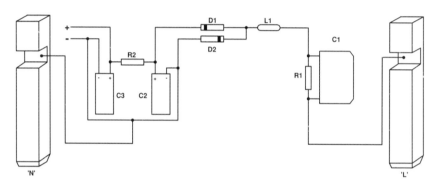

Figure 3.11 *Mains adaptor transmitter*

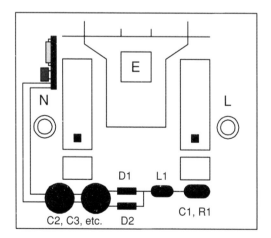

Figure 3.12 *Top view of a room transmitter built into a two-way mains adaptor*

Important notice

Electric shock

This and other mains powered devices can be lethal if incorrectly installed. For example, note that if the 'live' and 'neutral' wires were exchanged, then the casing of the microphone insert would then be live. If the capacitor C7 were to become short-circuit, then the entire circuit would be at mains potential. If any mains powered device is discovered, it should not be touched or dismantled until the unit is first isolated from the mains supply by a competent person.

Fire risk

Any device that uses either mains or battery power can be a potential fire hazard. High power batteries, if shorted out, may ignite combustible material, or cause burns to skin, etc. A circuit that provides a high RF level may use heatsinking on output transistors, etc. which may themselves cause burns to inquisitive fingers.

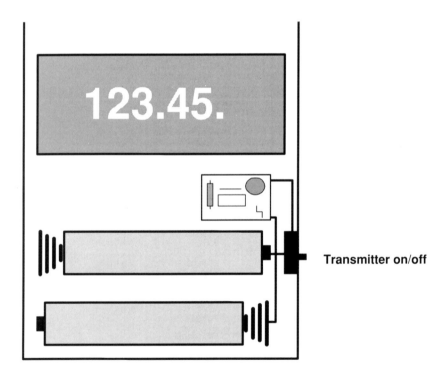

Figure 3.13   *Calculator transmitter*

Voice transmitters inside calculators and pens

With a little ingenuity, the device shown in Figure 3.5 may be built on a very small board and then hidden inside calculators and pens. If hidden inside a calculator, the 3 V supply for the calculator is usually utilized, being switched with a miniature slide switch to enable the transmitter to be turned on and off, as seen in Figure 3.13. If the circuit is built inside a pen, it is usual to either power the circuit with a row

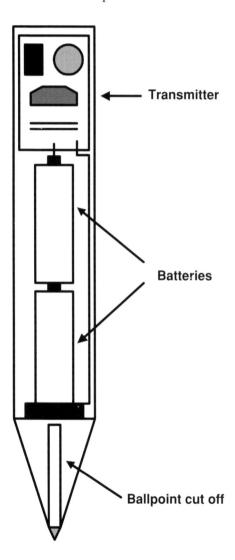

Figure 3.14 *Pen transmitter*

of watch/hearing aid button type batteries, or cylinder lithium batteries, as seen in Figure 3.14. In the case of the transmitter built inside a small space, it will be found that the 6 mm diameter tuning coil will have to be replaced with a coil consisting of around eight or nine turns, with a diameter of only 3 mm or so. Depending upon the circuit design and supply voltage, the transmission range will only be around 15 m, which is enough to transmit into an adjoining room. Whereas the length of the aerial wire inside a calculator transmitter can be equal to the outline of the calculator casing, the aerial wire of a pen transmitter will be shorter. Some companies supply crystal controlled transmitters inside two-way mains adaptors, pens and calculators, which, if the complexity of such circuits are considered, is a remarkable feat!

High power transmitters

The use of transmitters which have a more powerful output than the 'flea-power' transmitters mentioned elsewhere in this book are sometimes required when there are many obstacles in the path of the surveillance transmitter and monitoring station receiver, or the distance between them is too far so as to make a low powered device feasible. Whereas a typical microtransmitter will produce an RF power in the order of just a few milliwatts, i.e. a few thousandths of a watt, the VHF-FM transmitter described in Figure 3.15 has a power output of between around a half and 2 watts, depending on the power source, which may be anywhere between 6 volts and 30 volts d.c. The battery or batteries should be of the alkaline high power type, since the current drain will be found to be relatively higher when compared to microtransmitter current drain. The power output of this device is somewhat proportional to the current drain and so therefore both may be decreased by altering the value of R6 to a higher resistance, or a variable resistor with a value of around 1k may be introduced in series with the existing R6, so as to give a variable power output. The variable resistor must not be a wirewound device because this would act as an inductor which will cause feedback problems.

The audio input to the power oscillator, which incidentally is formed by TR2 and associated components, is derived from a piezoelectric microphone which drives the simple audio frequency amplifier TR1. The input of the audio amplifier is controlled by the gain pot R1, which selects the correct amount of voltage that is generated by the piezoelectric microphone, then connects this signal to the base of audio amplifier TR1 via C7.

It may be found that there is insufficient housing space for a bulky

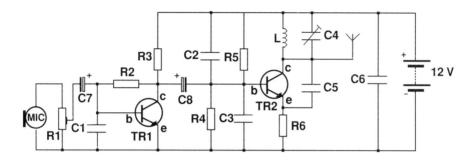

Component listing for 1 watt transmitter

Resistors
R1 = 27k
R2 = 330k
R3 = 5k6
R4, 5 = 10k
R6 = 100R

Semiconductors
TR1 = BC547
TR2 = 2N2219 fitted with heatsink
MIC = piezoelectric microphone

Capacitors
C1, 2, 3, 8 = 330 pF
C4 = 2–10 pF trimmer
C5 = 4p7
C6 = 1 nF

L = 6 turns 22 gauge enamelled wire wound on ³⁄₁₆" former

Figure 3.15 *High power transmitter, 1 watt*

piezoelectric microphone, so with a slight modification to the circuit, it is possible to employ an electret microphone insert as shown in Figure 3.16.

If the device is to be used as a telephone line monitor transmitting device, the interface circuit shown in Figure 3.17 may be used. Capacitors C9, 10, 11 and 12 are d.c. blocking/audio coupling devices. Resistors R7 and R8 are to attenuate the high audio levels that are present on the telephone line that would otherwise overdrive the audio and produce a very poor audio and RF signal. They also provide some degree of isolation between the telephone line and the transmitter. Diodes D1 and D2 will act as clipping diodes should anything go amiss by clipping the audio or voltage spikes from the telephone line to a maximum of 0.7 V.

Since the RF field that is generated by this transmitter is relatively large, the problem of RF feedback may very well be encountered. This may be overcome by placing the transmitter inside a metal enclosure, keeping all internal wiring as short as possible and the aerial wire

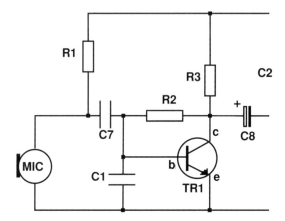

Component listing for electret microphone modification

MIC = electret microphone insert
C7 = 100 nF
R1 = 4k7

Figure 3.16   *Modification for using electret microphone*

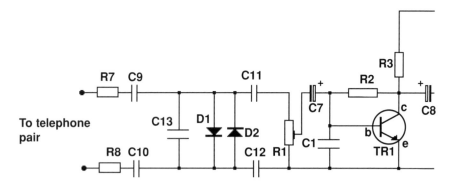

Component listing for phone to transmitter interface unit

R7, 8 = 10M
C9, 10, 11, 12 = 100 nF
C13 = 1 nF
D1, 2 = 1N4007

Figure 3.17   *Interface for telephone to 1 watt transmitter*

away from the power source. Suitable RF decoupling capacitors, whose values are in the range 470 pF to 1 nF can be used if and where required.

Theory of the 'Look – no batteries!' parasitic transmitter

Rumour had it that once upon a time in the midst of the cold war days, a transmitter was installed in the enemy headquarters that could be remotely switched on and off by the perpetrators, and that the hidden transmitters were not powered by mains supply or battery. The hidden transmitters were built into every wall in the embassy by local workers before they were inhabited by the diplomats and their co-workers. Sounds somewhat farfetched? Referring to Figure 3.18, it can be seen how this system can work. The surveillance centre would have a very powerful carrier-only transmitter close to the target area. This transmitter would then beam

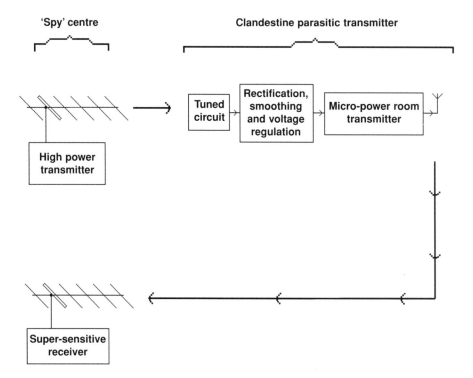

Figure 3.18   *Theory of parasitic 'no batteries' transmitter*

megawatts of RF power, via a highly directional high gain antenna at the hidden room transmitter which has a high-Q tuned circuit, tuned to the frequency of the high powered transmitter, which is followed by a series of standard power supply components, i.e. a rectifier, smoothing and voltage regulation section. The d.c. current derived from this circuitry is connected to a room transmitter that generates only a very small RF signal, which is then transmitted back to the monitoring centre, which is fitted out with a high gain directional receiving aerial and dedicated receiver. If this still sounds farfetched, check out the theory on how hidden clothing tags in stores sometimes work, or how the hidden strips inside books and video tapes also operate. Some intelligence agencies that first accidentally discovered large amounts of RF power in the embassy in which they worked originally believed that it was an enemy plot to 'microwave' their brains or whatever, before the hidden transmitters were finally discovered. Needless to say, when things got too hot for the surveillance centre, they would simply switch off the high power transmitter, which would mean that the hidden transmitter would stop transmitting immediately.

The 'Trojan' transmitter

Sometimes it is impossible to gain entry to enable the operative to plant a transmitting device anywhere in the target area for one reason or another, whether by fair means or foul. To this end, the idea of the Trojan transmitter was born. A transmitting device can be hidden inside an attractive present, a piece of office machinery, or for domestic purposes, a music centre, dressing/makeup table, etc. The Trojan is simply mailed to the target at home or at work, with 'free gift' stickers liberally attached, so by human nature the package will usually be eagerly accepted. In the past, discussion has often favoured no stickers, paperwork or anything. The target would in this case probably just put the unsolicited goods to one side for a few days in case of a follow-up bill, which when not received will give cause for the 'freebie' to be accepted gladly.

To give longevity to the transmitting device, and to make the system cost effective, a very high capacity battery or module of high capacity batteries connected in parallel so as to give a very long transmitting life span must be used. An alternative approach could be to use only a very low power transmitting device that would conserve limited battery power. However, if the Trojan is a large, moderately heavy 'horse', then the power source for the concealed transmitter could be a heavy duty sealed lead–acid battery as is used in burglar alarm control panels as a mains power failure backup power source. Other batteries could be rechargeable nickel–cadmium cells, but these tend not to have the excellent capacity of the former type.

Mains charging circuit for a Trojan transmitter

The joy of sending a horse that is mains operated is that it will sooner or later hopefully get plugged into a mains socket and switched on, perhaps left plugged in permanently. To this end, it has been thought prudent to include a sealed lead–acid battery mains charger circuit, as shown in Figure 3.19. Note that this circuit should only be used as either a standalone low current mains supply for some devices, or for charging only sealed lead–acid batteries as specified by their manufacturers' instructions, otherwise an explosion or fire hazard may occur if other battery types are used. The mains transformer should be good quality and mounted so that physical hum is kept to a minimum, since the target might become annoyed at the loud and worrying buzzing that will emanate from their free makeup table with pretty backlighting. The voltage regulator should be mounted on a sufficiently large heatsink, since although the transmitter will only take a small amount of current from the system, if the battery is finally

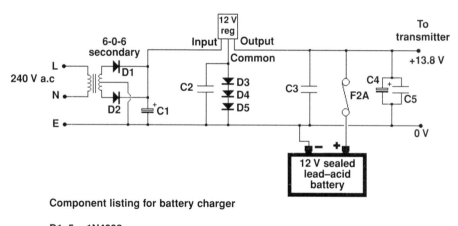

Component listing for battery charger

D1–5 = 1N4002
REG = LM7812
C1 = 1000 μF/25 V
C2 = 100 nF
C3, 5 = 1 nF
C4 = 100 μF/25 V

Figure 3.19 *Mains charging circuit for battery transmitter*

put onto charge after a long time when it is in a discharged state, the current drawn through the voltage regulator will be quite high. To this end, a mains transformer with a current rating of 2 amps should be used.

4 Telephone transmitters

As described previously, telephone transmitters fall into the two distinct categories of parallel and series devices. As in the case of room, or voice transmitters, they may be of either of the VFO or crystal controlled oscillator design. The sub-categories of telephone transmitters are parasitic, or leech types, that use a proportion of the power on telephone wiring, or are self-contained, i.e. they have their own battery or other power source. To appreciate how a parasitic device obtains power, the telephone system must be investigated. Figure 4.1 shows a representation of a standard 'master' telephone socket. We are not too concerned with the components that are found already existing within a master telephone socket, but the use of a 'surge arrestor' device, which is used to protect the system equipment from high voltage spikes caused by lightning, etc., is occasionally used in telephone surveillance designs. If a voltmeter were connected across the line wires, terminals 2 and 5, with the telephone handset 'on the hook', a normal voltage of around 50 V would be measured. If a meter for measuring current were inserted in series with one of the lines, it would be found that no current is flowing, since the circuit is open-circuit. If the telephone handset were lifted from the cradle, 'off

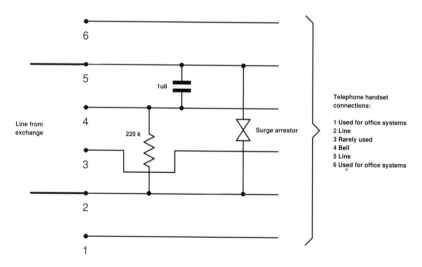

Figure 4.1 *Telephone master socket details*

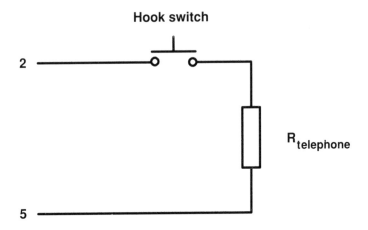

Figure 4.2a   *Series connection of telephone transmitter – 'off the hook'*

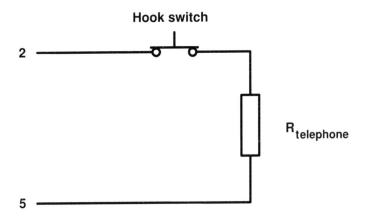

Figure 4.2b   *Series connection of telephone transmitter – 'on the hook'*

the hook', it would be noticed that current now flows, and the voltage across the telephone line will fall to around 10 V, depending on the telephone system being used. Following on from this, it can be appreciated that the telephone handset can be regarded as a simple resistor, with a series switch, normally off (the 'hook switch'), as seen in Figures 4.2a and 4.2b. If a further resistor were to be placed in series with the existing one (the telephone) without adversely affecting the correct operation of the telephone, by Ohm's Law, a voltage would appear across the freshly introduced resistance every time the handset were to be lifted, then disappear when the telephone handset was replaced, see Figures 4.3a and 4.3b.

The circuit designer will now have a supply voltage, varying in

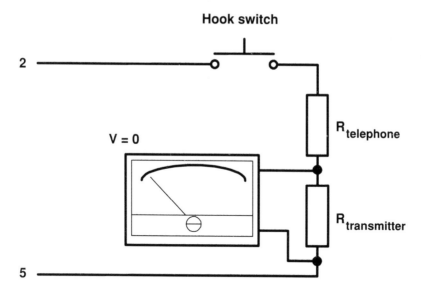

Figure 4.3a *Series connection with a further resistor fitted – 'off the hook'*

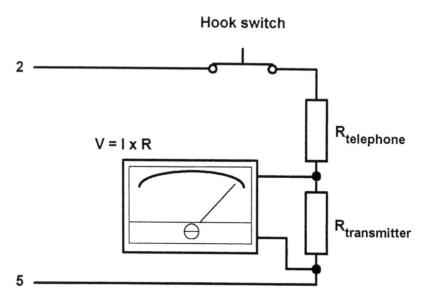

Figure 4.3b *Series connection with a further resistor fitted – 'on the hook'*

amplitude, and of uncertain polarity. To obtain a usable d.c. voltage (since the polarity of the voltage on the telephone line may swap over in various modes of use), all that is required is a simple bridge rectifier circuit, as seen in Figure 4.4, to ensure correct polarity at all times. By

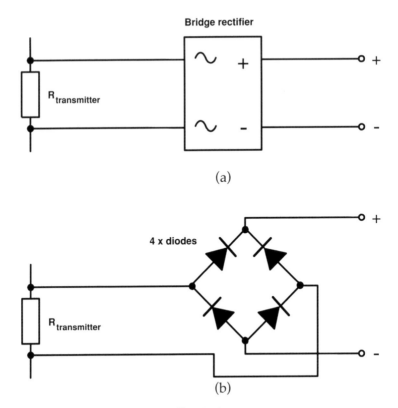

Figure 4.4a and b   *Simple bridge rectifier circuits*

using a bridge rectifier circuit, no matter which way around the device is connected to the telephone line, or regardless of the polarity of the voltage of the line, the output terminals of a bridge rectifier will be correct. Since the voltage on the line is d.c. when the telephone is actually in use, no smoothing is required in a low component count design. Any alternating voltage, or ripple, on the telephone line, is the audio signal, which is used to modulate the transmitter. Although a bridge rectifier device is shown in use in the following circuits, to keep construction small, four individual rectifiers are actually used in construction.

Simple series telephone transmitter

A simple series telephone transmitter is shown in Figure 4.5, with typical construction on stripboard shown in Figures 4.6a and 4.6b. The actual construction of the unit is not at all critical. By using small components, and soldering a few components underneath the board,

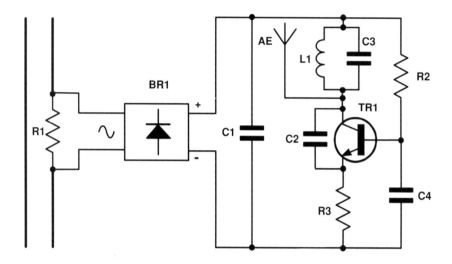

Figure 4.5 *Simple series telephone transmitter*

this circuit can be installed inside a two-way plug-in telephone adaptor with ease, although the aerial wire would only be a few centimetres in length.

Although only two breaks in the copper strips are shown, it may be prudent to cut any lengths of unused copper strip to prevent unwanted capacitive effects being introduced. Breaks in the copper strips can be made either by using a tool that is commercially

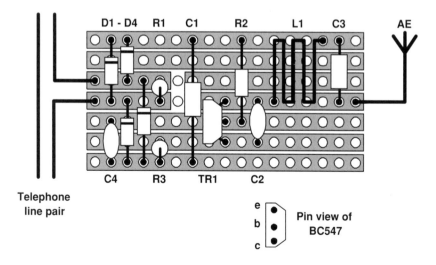

Figure 4.6a *Simple series telephone transmitter component layout*

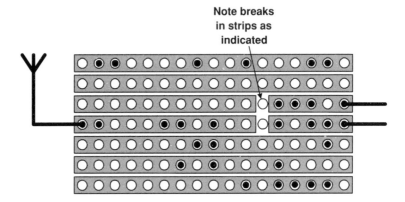

Figure 4.6b   *Simple series telephone transmitter – underside of stripboard*

available for this purpose, or by cutting the copper away with a small drill bit held with the fingers. With the exception of a few components, the circuit of the transmitter is the same as that of the simple voice transmitter in Figure 3.3. TR1 forms the oscillator. Modulation is derived from the audio on the telephone line, which is then superimposed on the supply voltage lines of the circuit.

Component listing for the simple series telephone transmitter, Figure 4.6a

Resistors

R1	330 R	R2	100 k	R3	330 R

Capacitors

C1	1 nF	C2	5 p6	C3	5 p6

Semiconductors
TR1   BC547, ZTX300, etc.
BR1   bridge rectifier made from 4 × 1N4148 diodes

Inductor
7 turns 22 swg, 6 mm diameter

To enable tuning of the transmitter frequency, the coil may be iron dust slug type, or C3 may be of the trimmer type. The aerial wire should not be more than 150 mm or so, since because there are no RF chokes in the supply of the circuit, feedback may cause the device to stop transmitting. The simple transmitting device may be greatly improved upon by adding RF filtering and improving the modulation

Telephone lines

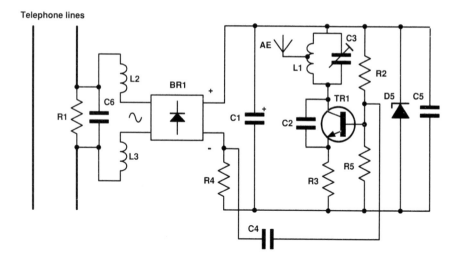

Figure 4.7 *Series telephone transmitter with RF filtering*

method, as seen in Figure 4.7. The two inductors, L1 and L2, along with capacitor C6, will stop RF re-entering the circuit. Resistor R4 and capacitor C4 improve audio modulation of the transmitter.

Component listing for telephone transmitter with filtering, Figure 4.7

Resistors

R1	680 R	R4	22 R
R2	22 k	R5	22 k
R3	330 R		

Capacitors

C1	10 μF	C4	4.7 μF
C2	5 p6	C5	1 nF
C3	5 p6	C6	0.1 μF

Inductors

L1	as Figure 4.5	L2, L3	10 μH

Semiconductors

D1, 2, 3, 4	1N4148
D5	6 V zener diode

A further improvement may be made to the circuit shown in Figure 4.7. For greater transmission range, a power amplifier/buffer may be added after the oscillator circuit, as shown in Figure 4.8. Two turns of

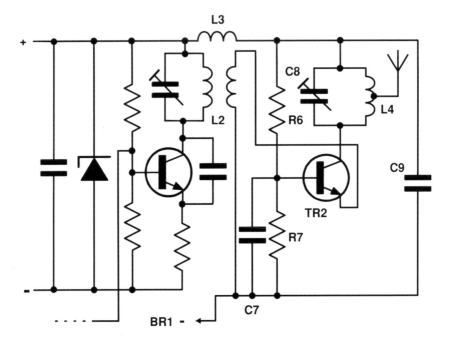

Figure 4.8   *RF filtered series telephone transmitter with RF amplifier*

enamelled wire are wound on top of the tank coil, to act as a secondary winding. This circuit may prove difficult to tune, since when C8 and L4 are tuned for maximum output, the frequency of the VFO will have to be readjusted to compensate. While the circuit in Figure 4.7 can have a transmission range of around 200 m, this circuit will have a much greater range.

Whenever designing a surveillance device for use on a telephone system, care must be taken when either drawing excessive current from the telephone system, or introducing too large a resistance in series with the existing equipment, so as not to cause a fault condition.

Component listing for higher powered series telephone transmitter, Figure 4.8

Resistors

R6	27 k	R7	4 k7

Capacitors

C7	10 nF	C9	10 nF
C8	5 p6		

Inductors

L2	two turns of enamelled 22 swg on top of tank coil

L3	4.7 μH
L4	as Figure 4.5

Semiconductor	
TR2	ZTX300, etc.

Simple parallel connected telephone transmitter

A simple parallel connected telephone transmitter circuit is shown in Figure 4.9. The circuit is almost identical to that of the simple series

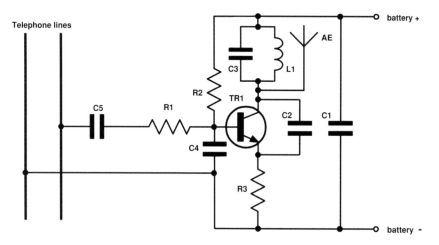

Figure 4.9 *Simple parallel telephone transmitter*

telephone transmitter shown in Figure 4.5. The circuit is powered by a PP3 9 V battery, and because the device uses a power source that is independent of the telephone line, RF filtering should not be required as in the case of a leech device, provided the aerial wire is not too close to the telephone feed wires. The amount of RF power produced by a self-powering unit is only dependent on circuit design and power source, therefore very high powered, long range devices are often available. With reference to the circuit shown in Figure 4.9, the capacitor C5, with a value of around 1 nF, blocks the d.c. from the telephone line but allows audio to reach the base of TR1. Resistor R1 limits the amount of audio to the base of TR1, and will typically be 100 k. This resistor should also protect the base of the transistor from high voltage spikes that appear on the telephone lines.

A slightly more complex parallel telephone transmitter, with better design regarding deviation and higher 'invisibility' to line checks is shown in Figure 4.11.

Telephone surveillance transmitters, if installed outdoors, can be rainproofed by enclosing the circuit board in a suitable plastic potting box, which is then filled with potting compound, epoxy resin glue or even car body filler for a cheaper alternative.

Component listing for parallel transmitter, Figure 4.9

Resistors

R1	100 k	R2	15 k	R3	220 R

Capacitors

C1	1 nF	C3	5 p6
C2	5 p6	C4	1 nF
		C5	1 nF

Inductors
As Figure 3.3

Semiconductor
ZTX300, BC109, etc.

A further circuit for a parallel telephone transmitter, using a FET device between the telephone line and oscillator, is shown in Figure 4.10.

The design in Figure 4.11 may be altered, by using component changes, to operate on either a 1.5 V or a 9 V supply. Component changes for the circuit using a 9 V supply are shown in brackets in the following components listing.

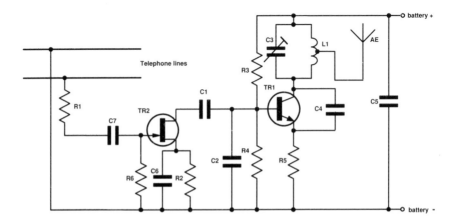

Figure 4.10   *Parallel telephone transmitter with FET input*

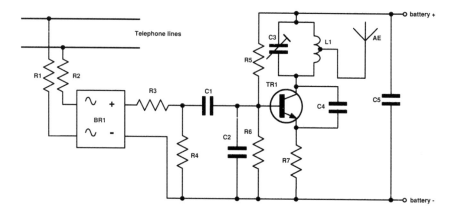

Figure 4.11   *Parallel telephone transmitter*

Component listing for parallel telephone transmitter, Figure 4.10

Resistors

R1	100 k	R6	2 M2
R2	270 R		

Capacitors

C6	1 μF

Semiconductors

TR2	2N3819 or equivalent

Other components as Figure 4.11 (9 V version).

Component listing for parallel telephone transmitter, Figure 4.11

Resistors

R1	10 k	R4	72 k (470 k)
R2	10 k	R5	15 k (27 k)
R3	470 k	R6	27 k
		R7	68 R (470 R)

Capacitors

C1	1 μF

Semiconductors

BR1	bridge rectifier

Other components as Figure 4.9.

5 Switching devices

Switching devices are frequently used in surveillance devices. They may be used for a variety of purposes, e.g. turning tape recorders on and off whenever a sound is registered, switching recorders on and off if a telephone handset is picked up and activating hidden radio transmitters or video recorders.

The first circuit to be considered is shown in Figure 5.1, which is designed to switch on a miniature transmitter whenever a sound is picked up within the range of the microphone. The circuit is capable of switching on any device, making that device voice activated, so long as the current requirements of the driven device do not exceed the current sourcing capability of the timer IC. If any more current is required, it would be necessary to add on a power transistor or relay. Sounds are picked up by the microphone and amplified by the transistor, which has two outputs from the collector. One output is fed to the base of the transmitter, point 'X' of Figure 5.2, and the other

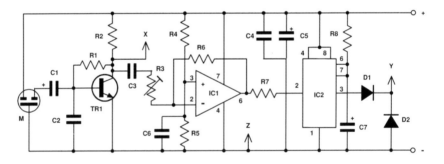

Figure 5.1 *Voice activated switch for transmitters*

output is fed to an op-amp (operational amplifier) IC, IC1, the amount of audio drive being attenuated by variable resistor R3. The output of the amplifier (whose overall gain is controlled by R6), is fed to the trigger of the timer IC, IC2. The output of the timer chip, pin 3, will now go high for a predetermined period of time, which is set by R8 and C7. The time is set for a delay before switch-off of around five seconds, but may be altered if R8 is made variable. The output, at pin 3, now becomes the supply voltage of the driven device, and in this case, is connected to 'Y' of the transmitter. Whenever deciding to use

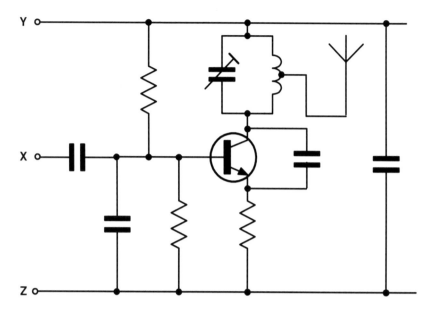

Figure 5.2 *Connection of transmitter to voice activated switch*

a voice activated switch, abbreviated to a 'VOX', it is best to weigh up the standby current of the switch along with the power used if a device were left to operate 'full-time' without switching.

Component listing for the VOX and audio amplifier, Figure 5.1

Resistors

R1	2 M2	R5	100 k
R2	15 k	R6	4 M7
R3	1 M pot	R7	100 k
R4	100 k	R8	100 k

Capacitors

C1	1 μF	C4	330 pF
C2	330 pF	C5	330 μF
C3	10 nF	C6	100 μF
		C7	47 μF

Semiconductors

TR1	BC547, etc.
IC1	741 op-amp
IC2	555 timer
D1, 2	1N4148

Microphone
3-pin electret microphone insert

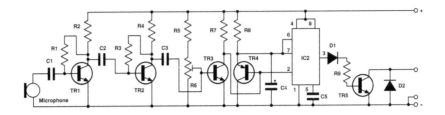

Figure 5.3a *Voice switch*

A similar circuit to the above can be seen in Figure 5.3a. The difference in this circuit is that the op-amp IC is replaced by a circuit using more audio amplification transistor stages. R8 and C4 control the hang time of the timer chip, whereas R6 controls the sensitivity of the device. Although the circuit uses a crystal microphone, with slight modifications, it may well work with selected dynamic or electret microphone inserts. Figure 5.3b shows how the current switching capabilities of this and similar circuits may be vastly increased by the use of a relay to switch heavy loads, as well as improving the isolation between circuits.

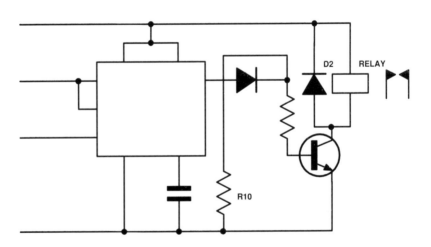

Figure 5.3b *Voice switch with modifications*

Component listing for the VOX in Figure 5.3a

Resistors

R1	2 M2	R6	10 k pot
R2	6 k8	R7	1 k
R3	2 M2	R8	150 k
R4	6 k8	R9	1 k
R5	100 k	R10 (Fig. 5.3b)	1 M

Capacitors

C1	100 nF	C3	100 nF
C2	100 nF	C4	100 μF
		C5	100 nF

Semiconductors

TR1, 2, 3, 5	BC547
TR4	BC327

Microphone
crystal insert

Telephone to tape recorder interfacing

Some typical telephone to tape recorder interfaces can be seen in Figures 5.4, 5.5, 5.6 and 5.7.

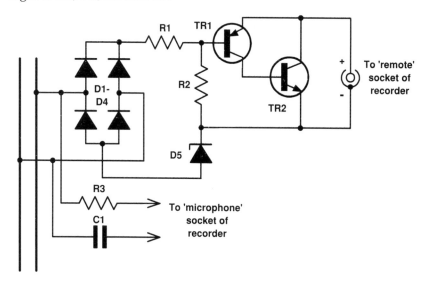

Figure 5.4 *Telephone to tape recorder interface*

Considering the circuit in Figure 5.4, the circuit is off, i.e. there is a high resistance presented to the remote control terminals of the cassette tape recorder, whilst the telephone handset is on the hook. This is because the zener diode is conducting because there is 50 V across the two lines. When the handset is lifted, the voltage across the lines will fall to around 10 V. This will have the effect of turning the zener diode off, turning on the PNP transistor, so switching on the NPN transistor. This has the effect of presenting a low impedance at the remote terminals of the recorder, so turning the recorder on. Audio is tapped from the telephone lines by R3 and C1, which are connected to the auxiliary, or external microphone input socket of the recorder. This circuit may sometimes produce motor noise which is recorded by some recorders, in which case, the points where R3 and C1 are connected to the lines may be reversed to cure this problem.

Component listing for interface circuit, Figure 5.4

Resistors
R1, 2, 3   33 k

Capacitor
C1   10 nF

Semiconductors
D1, 3, 4   1N4148
D5   15 V zener diode
TR1   BC327 PNP
TR2   BC547 NPN

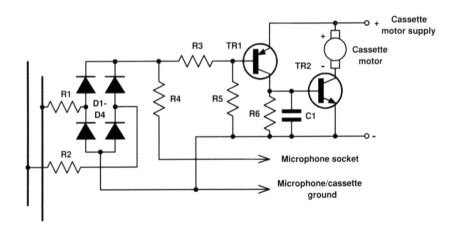

Figure 5.5 *Telephone to tape recorder interface*

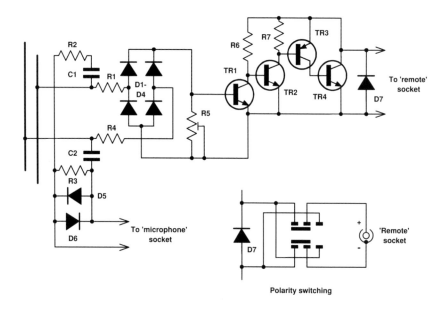

Figure 5.6   *Telephone to tape recorder interface with polarity switch*

The circuits in Figures 5.5 and 5.6 work on pretty much the same theme, that of using PNP and NPN transistors to provide the correct logic. The circuit in Figure 5.6 has a few refinements. Diodes D5 and D6 provide a clipping circuit so that the audio signal sent to the cassette recorder does not exceed 0.7 V. The inset of Figure 5.6 also shows a method of switching the polarity of the output terminals of the circuit. This is required because some recorders do have an opposite polarity with respect to the remote socket terminals.

Component listing for interface circuit, Figure 5.5

Resistors

R1	10 k	R4	22 k
R2	10 k	R5	330 k
R3	1 M	R6	27 k

Capacitor
C1   10 μF

Semiconductors

D1, 2, 3, 4	1N4148
TR1	BC327
TR2	BC547

Component listing for interface circuit, Figure 5.6

Resistors

R1	220 R	R4	4 M7
R2	22 k	R5	1 M
R3	4 M7	R6	56 k
		R7	100 k

Capacitors

C1, 2   100 nF

Semiconductors

D1–7	1N4148
TR1, 2, 4	BC547
TR3	BC327

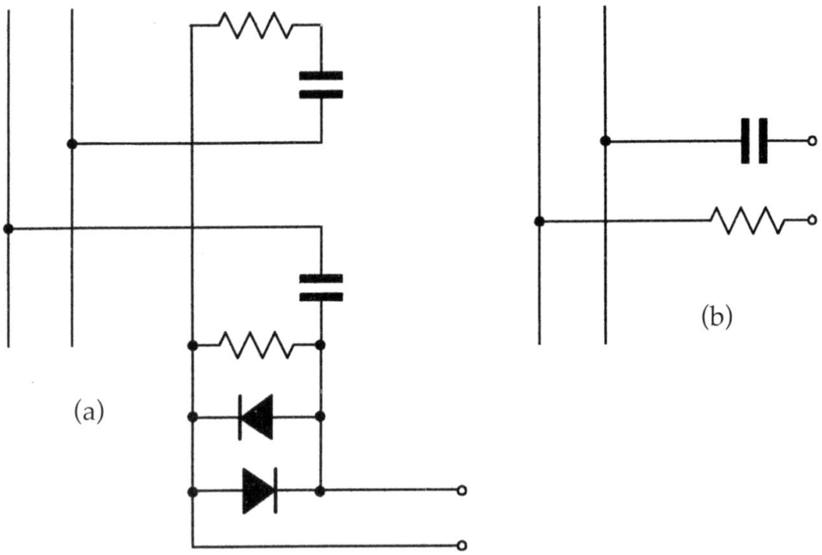

Figure 5.7a and b *Interfaces for telephone to VOX tape recorders*

If a voice operated tape recorder is available, then the circuits represented in Figure 5.7a and b may be used to interface a telephone to the recording machine. Either circuit can be encapsulated in a potting box, or if space permits, inside the recorder itself. The circuit in Figure 5.7b which consists of a 10 nF to 100 nF capacitor, and a 100 k resistor, can be built inside a standard 3.5 mm jack plug body, to be plugged into the recorder, with the other end of the lead terminated with an ordinary telephone plug. If any of these circuits are used, they may be simply plugged into a 'Y' two-way adaptor to connect with the telephone system, which allows the possibility of unplugging them in a second.

Conversion of radio/cassette recorders to 'RF VOX'

Depending on the type of unit available, it is possible to convert a standard radio cassette to become 'RF VOX' activated, i.e. when the receiver section picks up a signal from a switched radio transmitter, such as a voice activated transmitter, or a series leech telephone trans-mitter as described earlier, the tape recorder section of the unit will become activated, recording any conversations that are transmitted by the covert device.

To achieve the correct switching, it is required that a feed from the audio section of the receiver is able to turn the cassette recorder section on when an audio signal is present, and off when the audio signal ends. The circuit that fulfils these requirements is shown in Figure 5.8. Any audio signal that exceeds a pre-set level, controlled by R3, will turn on the cassette recorder motor. Power for the circuit is derived from the unit, and as in the preceding circuits, the ground return lead from the motor has been in effect 'cut' by the output transistor of the control circuit, in this case TR3. Regarding the circuit in Figure 5.8, C1 can be omitted if no adverse effect upon the audio output stage of the receiver is noted. It may also be possible to put C1 in the circuit and leave out R2 and R3 if there is a lot of audio signal available, connecting C1 directly to the base of TR1. Another modification is the inclusion of Rx, which is included in series with

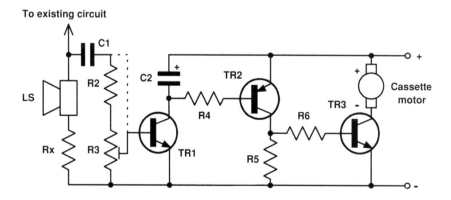

Figure 5.8 *Conversion of radio cassette to RF VOX*

the loudspeaker, if the volume of the loudspeaker has to be uncomfortably high in order to get the motor to be activated by the switching circuit. The value of Rx must be found by trial and error, since it depends on the unit being modified, and should be of a suitable wattage to be able to cope with the power dissipation, and may be shorted out with a switch, if required, so that if the recorder is used for playing back any recorded conversations, this may be done at full volume.

Output transistor TR2 must have sufficient current capability for the motor, and should be fitted with a heatsink if required.

Component listing for 'RF VOX', Figure 5.8

Resistors

Rx	see text	R4	100 k
R2	56 R	R5	100 k
R3	4 k7	R6	1 k

Capacitors

C1	10 μF	C2	10 μF

Semiconductors

TR1	BC327	TR2	BC547

Tape recorder speed control

It is often necessary to slow down the speed of a standard tape recorder so as to give extended recording time. There are several ways to slow down the speed of a cassette recorder motor. If a normal resistor is placed in series with the motor of a recorder, a maximum of around 15 per cent speed reduction will be possible, for after this, the initial current required by the motor for start-up will not be available, which will mean the recorder will not start 'from cold'. One method is to modify the mechanism of the unit. This is done as follows. Remove the pulley of the motor, slip the drive belt directly over the motor spindle, then replace the pulley, whole if space permits (or a cross section of the pulley if the casing of the recorder does not allow this), to prevent the belt from slipping off. A smaller belt may have to be obtained to obtain sufficient tension. The effect of making the diameter of the drive smaller will alter the gear ratio, so slowing down the motor. This method has the drawback that some cassette recorders have too much load on the belt, which will keep slipping on the smooth motor spindle. The majority of tape recorders already have an adjustable speed control system, which is altered by a pre-set potentiometer. The speed control resistor can be found on either the main circuit board, or in some cases, the motor body has a small rubber-covered access hole, in which a very small screwdriver is inserted to locate the pre-set. The speed of a cassette motor can be adjusted between 25 per cent and up to 500 per cent, with a large variation even between the same model.

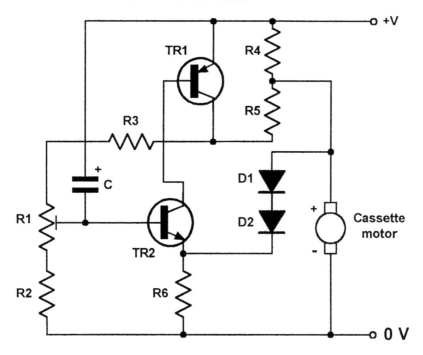

Figure 5.9   *Cassette recorder speed controller*

A typical circuit for adjusting the motor speed is shown in Figure 5.9. In Figure 5.9, a PNP transistor, TR1, acts as a series regulator, i.e. a 'variable resistance' in series with the motor and the supply line. If the voltage on the base is made more negative with respect to the emitter, it is turned on more and will supply more current to the motor via R5. When the variable resistor R1 is altered, this will affect the bias of TR2, which will in turn affect TR1. Feedback in the circuit maintains a current regulated supply to the motor. Capacitor C has the function of ensuring that on initial power-up, the capacitor will seem to be a short-circuit, thereby making sure that TR1 and TR2 are fully on and supplying maximum power to the motor, which it needs to start up. Power for the circuit is taken from the recorders' own supply.

To install the circuit, both the ground and positive supply to the motor are located, the latter being unsoldered. The supply lead is connected to the positive supply of the speed control circuit, and the ground of the circuit is then connected to the ground of the recorder. The flying lead from the motor is now connected to the junction of R5 and the anode of D5. The circuit may give around one-third speed, so a 'C-120' tape can give a recording time of 2×3 hours, a total of six hours recording time!

Component listing for cassette motor speed controller, Figure 5.9

Resistors

R1	470 R	R4	1 k
R2	1 k8	R5	2R7 (4 × 10 R
R3	1 k		in parallel)
		R6	330 R

Capacitor
C   4 µ7

Semiconductors

TR1	BC327	TR2	BC547 (see text)
D1, 2	1N4148		

Radio controlled remote switching

Remote control devices can be used in electronic surveillance, for remote switching of transmitters, recorders, etc. To provide a useful system, the radio transmitter should be modulated with an audio signal. This modulated signal is then picked up by a suitable receiver, which then decodes the audio signal into a set of instructions that can turn a device on, etc. A simple tone modulated transmitter is shown in Figure 5.10. An RF oscillator, as used in the design in Figure 4.5, the simple telephone transmitter, is modulated by the audio oscillator that is formed by IC1. The amount of modulation is controlled by R4, and the frequency/waveform of the audio signal is set by R2, R3 and C2. Note that this circuit was originally designed to act as a radio link in a simple intruder alarm. Transistor TR1 is switched on when a normally open latched switch is activated, supplying current through TR1 to power the transmitter and modulator. The transistor could be dispensed with, and a standard switch could be connected between

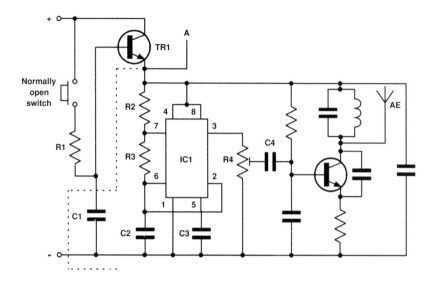

Figure 5.10 *Tone modulated transmitter*

the positive supply and point 'A'. If this is done, then a large decoupling capacitor, C1, should be connected between point A and ground.

A tone modulated transmitter would not be of much use without a tone decoder, and such a circuit can be seen in Figure 5.11. A standard radio receiver is used to receive the transmission, and the tone decoder is plugged into the earpiece socket of the receiver, or connected to any other suitable take-off point. The audio signal is fed

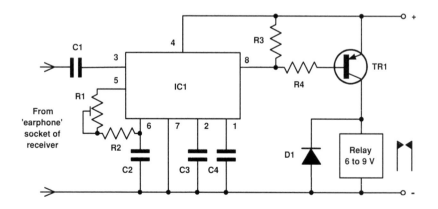

Figure 5.11 *Tone decoder*

to the input of the PLL integrated circuit. If the circuit is tuned to the same audio frequency as the transmitted audio frequency, by means of R1, then the output of the IC will cause the relay to be activated. These contacts could be connected so as to turn on another transmitter, tape recorder, close garage doors, etc.

Component listing for the tone modulated transmitter, Figure 5.10

Resistors

R1	10 k	R3	1 k
R2	4 k7	R4	470 R

Capacitors

C1	100 μF	C3	100 nF
C2	100 nF	C4	100 nF

Semiconductors

TR1	BC547
IC1	555 timer

Note: remainder of circuit as Figure 4.5.

Component listing for the tone decoder, Figure 5.11

Resistors

R1	22 k multi-turn	R3	22 k
R2	6 k8	R4	1 k

Capacitors

C1	10 nF	C3	2 μ2
C2	100 nF	C4	4 μ7

Semiconductors

TR1	BC327
IC1	567
D1	1N4007

Relay
150R 6–9 V

6 Video devices

Video systems

Electronic methods of information acquisition may be used that include the use of video systems and audio systems. The use of video techniques has been around in the form of security and surveillance systems for a number of years. Video cameras may be fixed upon a certain area, or can be mounted on an az-el (azimuth-elevation) remote controlled motorized mounting. Az-el mountings allow the camera to be turned around to produce a viewing field of up to 360 degrees, if mounted on top of a pole. More expensive systems will include an infrared light source, so that the camera can 'see' in the dark. Other systems include an audio channel, fed by a microphone and variable gain audio amplifier, enabling the security staff to listen in on possible breaches of security. The audio path can also be two-way, so enabling security personnel to talk back to the unsuspecting person.

The basic units required for remote surveillance regarding both audio and video are shown in Figure 6.1. The standard signal output

Camera

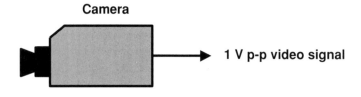

1 V p-p video signal

Microphone

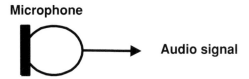

Audio signal

Figure 6.1 *Basic units for sound and vision surveillance*

from a video camera is a one volt peak to peak (1 V p-p) amplitude modulated waveform, but there is no standard for the output of a microphone since the impedance and voltage of the output will vary from one device to another. The video signal will often be required to travel the distance between the video camera and the security office monitor. For relatively short distances, e.g. 300–400 m, coaxial cable may be used. The distance will depend on the quality of the coaxial cable used, since the cheaper, poorer quality cable will have a greater resistance per length, thereby attenuating the baseband video signal so as to make it unusable.

Table 6.1   *Examples of attenuation in cables at 10 MHz*

Cable type	Approx. attenuation (dB/33 m)
UR90	1.1
UR57	0.6
UR77	0.3

Any mismatch in impedance will cause ghosting effects on the delivered picture, therefore it is important that the correct impedance is observed. The impedance found in most systems is either 50 ohms or 75 ohms. An alternative, but more costly medium for transmission, is that of fibre-optic cable, as shown in Figure 6.2. In this system, the video (or audio) signal modulates a light source at one end of the

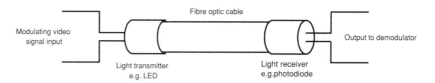

Figure 6.2   *Fibre-optic cable transmission*

cable. The modulated, 'flickering', light then passes down the cable, bouncing inside the internal wall of the fibre, being contained therein, until it reaches the other end of the cable, where the light falls on the surface of a detector, such as a photodiode. The varying light will now become a varying voltage or current that can then be processed to obtain the original information. If the distance between transmitter and receiver is too great to receive a good signal through co-axial cable, the baseband video signal may be boosted by using line driver amplifiers or other devices, as shown in Figure 6.3. Greater ranges of

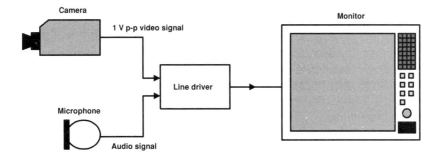

Figure 6.3 *Simple hard-wired system*

a few kilometres may be achieved by using a VHF carrier to transport the video signal through the long run of cable, which may need amplifiers along the length.

The most cost-effective method of sending a video signal to a remote monitoring station is by using a low powered transmitter, as shown in the simplified diagram in Figure 6.4. The type of circuit required to transmit a video signal need not be too complex, but can be if an audio signal is also required to be sent on the same carrier. Often seen on the market are so-called 'video senders', which are simple, low powered transmitters designed to transmit a signal, both video and audio, throughout a building, to enable the occupiers to watch a film being played on a video in another room. Anecdotes are often heard about such transmitters being used by individuals, who do not realize that the whole street has been receiving the wedding video through their television sets. This amusing anecdote does have a serious note, i.e. any video signal transmitted on an RF carrier could be intercepted, or interfered with, either accidentally or purposely, whereas a signal that uses a cable as a medium is less prone to these problems.

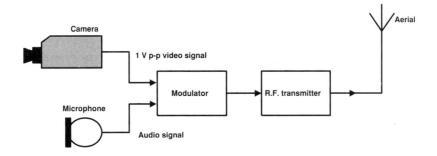

Figure 6.4 *Simple audio and video transmitter*

Camera types

The two main types of camera are the Videcon camera tube, as seen in Figure 6.5, and the charge-coupled device camera, or CCD, as seen in Figure 6.6. The operation of both types of camera are briefly outlined below.

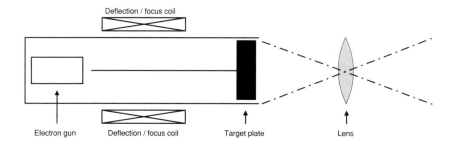

Figure 6.5   *Basic Videcon camera tube*

The Videcon camera tube has an electron gun in the base, which requires a heating element to produce loose electrons. These electrons are directed to a target plate by using deflection/focus coils. The target plate is covered in a material which has the property of photoconductivity. This plate is scanned in such a way as to provide an electrical output, the video signal. If the camera is to be used in low levels of lighting, an array of photodiodes would be used as the target plate.

In the charge coupled device (CCD camera), each individual pixel (picture element), is a separate photoconductive diode, laid down in a matrix. The picture required is focused onto the matrix by a lens

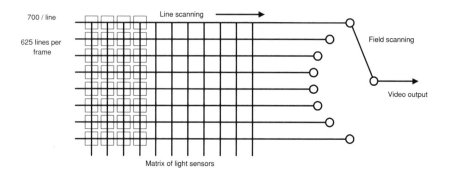

Figure 6.6   *Basic CCD camera*

system in the normal way. Each row of horizontal cells are read in turn, each cell in the row passing on information to its neighbour, in the same fashion as a row of firefighters would pass the bucket of water along the line. Once the information on one of the horizontal lines of cells has been passed on, the next line is read, and so on until a complete field, or frame, of information, has been read.

A basic video signal contains information on the following:

- The luminance level, i.e. brightness or darkness.
- A synchronizing signal that tells the monitor when a new line of information is about to be sent to it. This is called the line sync pulse.
- A synchronizing signal that tells the monitor when a new frame, or field, is about to be sent.

The obvious advantage the CCD has when compared to the standard tube is that since an electron gun, which requires a heating element, is not used, the power requirements are relatively minimal. The CCD camera being smaller, easily concealed, and powered by small batteries is ideal for portable or clandestine uses.

From Figures 6.7 and 6.8, it can be seen how an electronic waveform can be obtained by scanning a target, and how a picture can be rebuilt

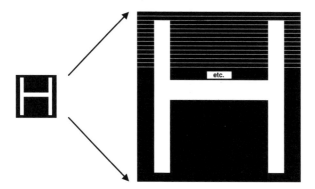

Figure 6.7 *Horizontal scanning by a television camera*

by a receiver. This example only shows two levels of luminance, those of peak black and peak white. Figure 6.8 demonstrates a waveform derived from scanning a line of bars that range from black, through the grey scale, to white. Since any monitor radiates a small amount of line, field and video signal, there is a slight danger of information being taken from computer systems by an operative picking up the signals from an insufficiently screened monitor, then 'rebuilding' the

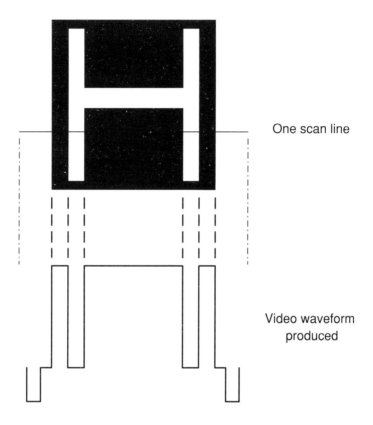

One scan line

Video waveform
produced

Figure 6.8 *Video waveform from one scan line*

screen information on their own monitor. This problem may be minimized by ensuring all cables are screened and earthed correctly, and if the monitor casing can be shielded, even better.

A low-powered video transmitter

The circuit of a low-powered (150 mW) UHF television transmitter is shown in Figures 6.9a and 6.9b. The standard video signal, 1 V peak to peak, is fed to the base of TR1. The resulting signal at point 'A' is clamped so the line sync pulse is at around 5 V. The video level at the input of the circuit is controlled by R1, and the biasing/clamping level is controlled by R4. Looking now at section two of the transmitter, TR3 is a crystal controlled oscillator, operating at around 120 MHz, depending on the frequency output required. TR4 and L5/L6 act as a frequency doubler and filter circuit. The resulting frequency of

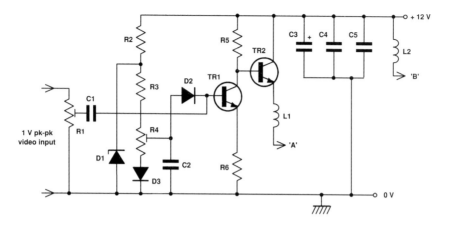

Figure 6.9a *Television transmitter – section 1*

around 240 MHz is doubled again by TR5, with L7 and L8 again acting as a bandpass filter. The final frequency of around 480 MHz is now fed to the final RF amplifier stage, TR6. The supply rail for the final 480 MHz stages is derived from the output of TR2, via the RF choke, L1, and is therefore modulated by the amplitude of the video signal. The supply to the crystal oscillator is regulated by IC1 to ensure a clean and stable signal.

For calculating the output, e.g. 470 MHz, the final frequency is divided by four, so in this example, the crystal required would be 117.5 MHz. Note that any deviation from this frequency would mean adjusting the value of all tuning capacitors and coils. Using only a telescopic whip aerial, a range of several hundred metres can be achieved with such a transmitter, and a range of a few kilometres or more if directional aerial arrays are used for both transmitter and receiver.

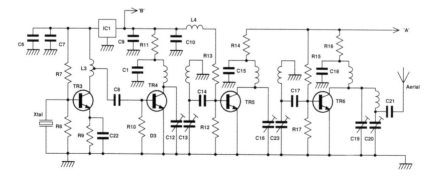

Figure 6.9b *Television transmitter – section 2*

Component listing for the low-powered video transmitter, Figures 6.9a and b

Resistors

R1	150 R	R9	330 R
R2	1 k	R10	15 k
R3	1 k8	R11	100 R
R4	470 R	R12	2 k2
R5	1 k	R13	22 k
R6	100 R	R14	100 R
R7	10 k	R15	10 k
R8	2 k2	R16	100 R
		R17	2 k7

Capacitors

C1	100 μF	C12	22 pF
C2	100 nF	C13	22 pF
C3	100 μF	C14	5 p6
C4	100 nF	C15	220 pF
C5	1 nF	C16	5 pF
C6	1 μF	C17	5 p6
C7	1 nF	C18	100 pF
C8	5 p6	C19	5 pF
C9	10 nF	C20	10 pF
C10	47 nF	C21	1 nF
C11	1 nF	C22	15 pF
		C23	5 pF

Semiconductors

TR1	2N2222	TR4	BFY90
TR2	2N2219	TR5	BFR34A
TR3	BFY90	TR6	BFR96
D1	5 V1 zener	D2	1N4148
D3	1N4148	IC1	8 V regulator

Inductors

L1	8 turns 26 swg 3 mm dia.
L2	10 turns 26 swg 3 mm dia.
L3	10 turns 26 swg 4.5 mm coil with core, tapped 3 turns from supply end
L4	3 turns ins wire on ferrite bead
L5, 6	2 turns 18 swg 5 mm dia.
L7, 8	1 turn 18 swg 8 mm dia.
L9	3 turns 26 swg 3 mm dia.
L10	2 turns 18 swg 5 mm dia.

Crystal	See text

Since the circuit operates at UHF, layout of the circuit is somewhat critical, requiring construction on a double-sided pcb (printed circuit board), paying attention to RF decoupling. Note that the coupling coils need to be close together in the actual construction of the circuit.

Automatic surveillance camera VCR switching

It is a very useful idea to have some method of being able to switch a video recorder on if a protected area has been invaded. This will mean that the video recorder will only operate if there has been some kind of violation, which means that valuable tape space, and the viewing of the tape by the security operative at a later date, is then minimized. The easiest method of performing this task is to have a system of commercially available PIR (passive infrared) sensors that cover the same area covered by the surveillance camera.

Figure 6.10 shows a typical basic arrangement, and shows how the units can be wired up together. The PIR sensor is mounted in such a way that if an intruder comes into range of the surveillance camera, the internal switch of the PIR is then closed. This normally open pair of contacts will control the video recorder via a suitable VCR

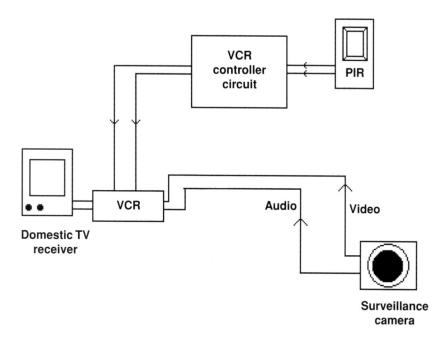

Figure 6.10 *Auto surveillance camera VCR switching*

controller circuit. Because the PIR has an integral timer circuit, the contacts will only be operational for the time period set, therefore the video recorder will be switched off after the preset period of time, unless the protected area is still occupied by a detectable body.

A simplified diagram of a VCR control switching unit is shown in Figure 6.11. A double pole relay is connected in such a way so that once the coil is energized via the PIR internal switch and a suitable low voltage supply, sets of contacts which are wired in parallel across the 'record' switch of the VCR will cause the recorder to go into the record mode. A further pair of relay contacts, which are normally open when the PIR relay and the controller relay are not powered up, will be connected in parallel with the 'stop' switch of the VCR, so stopping the recording when the PIR is not triggered.

If a particular VCR models seems incompatible with the idea of having the record switch held down for any length of time, it is possible to use a more complex circuit that uses a 555 timer as a one pulse device so that if a signal is received from the PIR, the 555 will close a relay for a second or two, so operating the recorder.

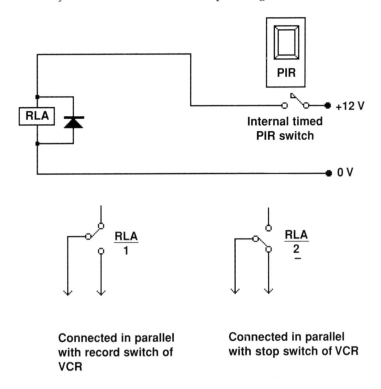

Figure 6.11 *Auto VCR switching diagram*

Vision switching circuit

The circuit of a four-way vision switching circuit is shown in Figure 6.12. The circuit in this case has four inputs that are standard 1 V p-p signals from four separate cameras. By the use of logic gates, any of the video inputs can be selected and viewed on a suitable monitor. Although this circuit uses a number of gates and devices to provide switching, it would be possible to make the circuit switch over using a timer circuit or even have a computer to do the switching automatically.

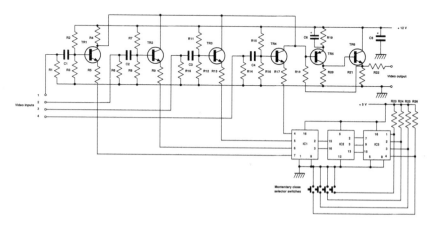

Figure 6.12 *Four-way vision switch*

Component listing for the video switching circuit, Figure 6.12

Resistors

R1	75 R	R9	390 R	R17	390 R	R25	1 k
R2	2 k7	R10	75 R	R18	1 k	R26	1 k
R3	1 k	R11	2 k7	R19	680 R		
R4	470 R	R12	1 k	R20	1 k		
R5	390 R	R13	390 R	R21	1 k		
R6	75 R	R14	75 R	R22	75 R		
R7	2 k7	R15	2 k7	R23	1 k		
R8	1 k	R16	1 k	R24	1 k		

Capacitors

C1	47 μF	C4	47 μF
C2	47 μF	C5	1 000 μF
C3	47 μF	C6	47 μF

Semiconductors

TR1	BC547	TR4	BC547
TR2	BC547	TR5	BC327
TR3	BC547	TR6	BC547
IC1	74139	IC2	7475
		IC3	74148

Switches   4 off momentary close

In this circuit, one of the video inputs is selected by IC1 effectively grounding the emitter of the chosen transistor, TR1, TR2, TR3 or TR4. If so required, the logic switching may be replaced by some kind of mechanical switching.

7 Countermeasures

There are several measures that may be taken by a person to avoid being the victim of electronic eavesdropping and surveillance methods. If anyone thinks that they are the victim of any type of 'electronic intrusion' by an individual or organization, then steps may be taken to minimize or if possible, exclude, any use of communication that may be intercepted and used against them. Apart from never communicating with another living sole ever again, there is no other way to obtain a 100 per cent secure situation. This is obviously a ridiculous situation. The level of any measures that are to be taken by a person must be justified by asking themselves the questions who and why – who would want to keep an interested eye upon them and why should the inquisitive party have any interest in them.

The more that a person becomes aware of the vast range of surveillance equipment that has been designed, then the greater are the chances of a type of paranoia setting in. At the other end of the scale, a private individual will ask him- or herself the question of who would possibly want to listen into a boring private conversation, surely the private conversations of people who are in the blaze of the public eye are the only ones that are of any interest to those involved in electronic eavesdropping. This is an easy trap to fall into, because when considering the number of private individuals who probably sit, for many a long hour, with their ears glued to their scanning radio receivers, which are capable of listening in to private telephone calls that use radio waves as a medium, maybe this is not so paranoid as first meets the eye. The world has moved on a long way from the original meaning of the word 'eavesdropping' (and the inverted drinking glass against a shared wall). Technology has now made eavesdropping possibilities an infinite and global affair, due to new communications methods.

Methods of countermeasures

The very first course of action to be taken when checking premises for hidden devices is a very thorough visual inspection of every square centimetre of surface area. Many films have depicted the suspicious person, who believes that the room or office they occupy has a concealed listening device, proceeding to rip up every article in the area, floorboards, chairs, pictures out of their frames, etc. only to find a miniature transmitter hidden inside a light bulb. This method of visual inspection is excellent and very thorough, but not very sociably acceptable. A visual inspection should be to look for any item that is obviously out of place in the area, such as wires trailing from the telephone system that are not part of the rest of the tidily installed cables and sockets. Dirty fingermarks around wall sockets, or damage to wall coverings around mains and telephone sockets may be suspicious. Once an area has been visually inspected, it may be the case that an offending device has been recovered, such as a disposable room transmitter fixed under a desk with adhesive tape. This may just be a decoy device, planted in such a fashion that discovery of the device is inevitable. Such a blatant tactic may be successful, but a professional will then start looking for the other devices that are hidden on the premises! There are many devices on the market designed to find out if electronic surveillance devices are in use, and some of the more common types of detectors and detection methods are now described.

Radio transmitter detection

A radio transmitter detector suitable for the detection of covert devices should be comprised of the following sections:

Aerial and tuned circuits

The tuned circuit should ideally be sharply resonant at the same frequency as the transmitter so that the maximum sensitivity to the wanted frequency is obtained, and which at the same time should offer a high rejection of unwanted radio frequency signals, such as those from commercial transmitters. In a design, it is possible to use a range of plug-in, interchangeable coils to cover a wide range of frequencies but still with an acceptable sensitivity. It is also possible to use a wide band 'front-end' with varying degrees of success. A professional design will have a front end that can be electronically switched throughout all the bands, in a similar way to the switching in scanning radio receivers, that automatically switch from one band to another whenever the end of a particular band is reached.

Output indication

The tuned circuit is fed to a circuit that will give an indication, by some means or another, that an RF field has been detected. This output may take many forms, and typical examples are:

1 A simple LED indicator that becomes illuminated, or changes colour, giving a visual indication.
2 A buzzer that gives an audible indication.
3 A trembler device that vibrates silently and secretly in a pocket or the hand, to warn the wearer that an RF field is close by.
4 A bargraph display to give an indication of field strength.
5 An analogue meter to give an indication of field strength.
6 A digital display giving a direct frequency readout of the hidden transmitter.
7 An audio output section that can produce a feedback howl between the loudspeaker of the detector and microphone of the audio transmitter.
8 An audio output section that can be listened to with headphones, so transmissions from the hidden transmitting device can be monitored.

Note that some of the indicators in the above list will enable the seeker to track down, or monitor, the hidden device, without actually alerting the operative who has planted the device, to this fact. This can often be useful in that it provides an opportunity to produce a certain amount of 'mis-information'!

Practical examples of radio frequency detectors

Some examples of practical RF detectors will now be discussed. For many years, the radio amateur has used 'field strength meters' for testing the output of amateur radio transmitters. These devices are very simple in their construction, and allow not only an indication that a transmitter is actually transmitting, but allow the relative field strengths, such as those around directional transmitting aerials, to be crudely measured.

A basic field strength meter is shown in both block and circuit diagrams in Figure 7.1. The aerial will preferably be a sturdy telescopic rod, the longer the better. The aerial is connected to the top of a parallel tuned circuit, which will be resonant (roughly) at the same frequency as the transmitter. If the design were for use in an amateur radio setup, then the output frequency of the transmitter would be a known factor. This would mean that the tuned circuit could be sharply resonant at the known frequency. Since the design of the field strength meter is for tracking down covert surveillance

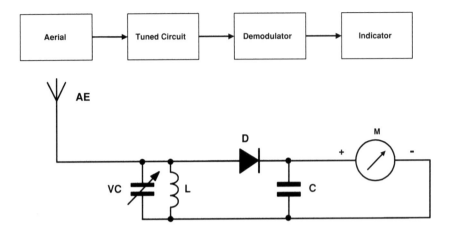

Figure 7.1 *Basic field strength meter*

transmitters, which by their very nature do not have a known transmission frequency, it is necessary that the bandwidth of the tuned circuit be as wide as possible, without sacrificing the sensitivity of the circuit too much. The diode D is a germanium signal diode, such as the OA91 device, which acts as a half wave rectifier, to recover the d.c. The capacitor C will act as a filter and will have a typical value of around 47 nF. The recovered and filtered d.c. is now fed into the

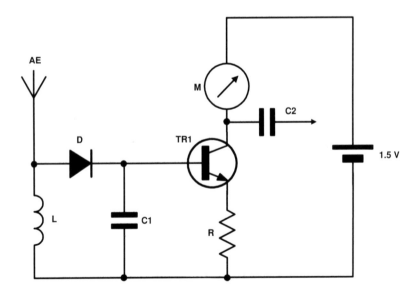

Figure 7.2 *Transistorized field strength meter*

meter M which is a sensitive ammeter with an fsd (full scale deflection) of only 100 µA or so. The circuit will be recognized as extremely similar to the old fashioned 'cats' whisker radio' of old.

Although the basic field strength meter will suffice in some situations, it is preferable to make the device much more sensitive, so that the tracking device can find very low powered transmitters. To amplify the signal that is detected, a circuit such as that shown in Figure 7.2 may be used. This circuit uses a transistor to amplify the recovered d.c., then feed this signal to a meter. Capacitor C2 has been included as an experiment to connect this circuit, via C2, to an audio amplifier with high gain. Hopefully, if the detected signal is strong enough, and contains amplitude modulation, then audio recovered may be heard. The circuit shown in Figure 7.3 has the advantage of having a sensitivity control, which will allow the sensitivity of the circuit to be lowered, a very useful feature when trying to actually pinpoint a hidden transmitter. In this design, the rfc (radio frequency choke) is chosen for operation at VHF. If it is preferable to give high frequency coverage, then an rfc with a value of around 1 mH would be substituted, or switched, into the circuit. The variable resistor, R2, doubles up as both a set zero and sensitivity control. The choice of the meter to be used in the circuit will also have a bearing on the sensitivity of the finished unit, as a lower fsd meter will give greater sensitivity.

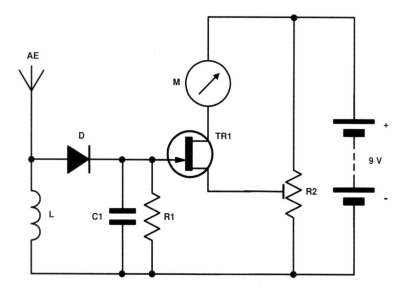

Figure 7.3 *FET field strength meter*

Component listing for the FET field strength meter, Figure 7.3

Resistors

R1	2 M2	R2	5 k pot

Capacitor

C1   150 pF

Semiconductor

TR1   2N3819

Inductor

L   RFC 10 μH

Meter, 25 μA to 1 mA, see text

A unit that can be used as both a field strength meter and a direction finder device is shown in Figure 7.4a. Once again, a tuned circuit and simple diode detector are used. The aerial is coupled to the coil so as

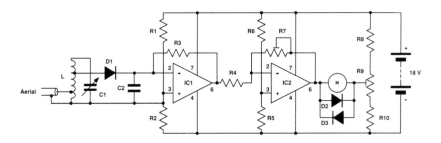

Figure 7.4a   *Field strength/direction finder unit*

to provide a good impedance match. The coil, or inductor, shown in Figure 7.4b, is fabricated from a 100 mm length of 18 gauge tinned copper wire. Taps for the aerial connection and for the detector diode are made at the points shown, with the length of wire then being bent up into an hairpin configuration, with a 10 mm radius. The signal from the detector circuit is connected to the inverting input of an op-amp, whose gain has been set by resistor R3. The output of the first op-amp is connected to a second op-amp device via R4. The meter, M, will give an indication of the strength of the signal received by the unit. Diodes D2/D3 are simply protection diodes, which ensure that a voltage of no greater than 0.7 V appears across the meter in either

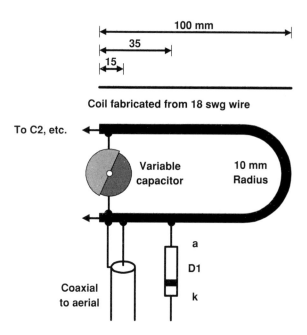

Coil fabricated from 18 swg wire

Figure 7.4b *Field strength/direction finder unit tuned circuit construction details*

direction. An attenuator circuit, that can be used between an aerial and a receiver, is shown in Figure 7.5. This is essential when trying to track down hidden high powered transmitters, or any transmitter in close proximity that may overload a tracking receiver. The aerial connection on the field strength meter should be a good match for any aerial with a 50 ohm impedance. When used in direction finding of

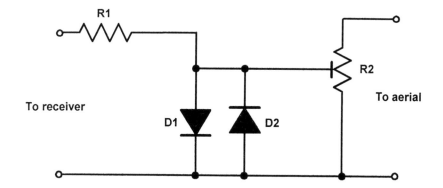

Figure 7.5 *Variable attenuator*

high powered transmitters, a suitable portable directional beam aerial should be used.

Component listing for the field strength meter/direction finder, Figure 7.4a

Resistors

R1	10 k	R6	10 k
R2	10 k	R7	1 M pot
R3	100 k	R8	2 k7
R4	2 k7	R9	470 R pot
R5	10 k	R10	2 k7

Capacitors

C1	30 pF variable
C2	1 nF

Semiconductors

IC1, 2	741 op-amp
D1, 2, 3	OA91

Inductor	see text and diagram

Component listing for the receiver attenuator

Resistors

R1	50 R
R2	100 R carbon pot

Semiconductors

D1, 2	1N4148

A hidden transmitter detector that can easily be built into a small space, with a visual (LED and/or meter) indicator is shown in Figure 7.6. The circuit is rather similar to the previous detector, but the integrated circuit chosen does not require a split voltage supply, and the output voltage can provide a full swing that goes practically from ground potential up to supply voltage. Although the circuit shows the use of a meter to compliment the LED visual indicator, the meter, and R6, may be omitted, enabling the circuit to be hidden, worn, etc. with just the LED showing. Although the circuit uses a tuned circuit, centring on about 120 MHz, the circuit may be modified by using a broadband front end such as that shown in the circuit for the FET field strength meter, Figure 7.3. The inductor is an air spaced coil, with around 4–5 turns of 22 swg enamelled wire, with an inside diameter of 6 mm. Although R3 may be omitted, it is used as 'fine tuning' to set

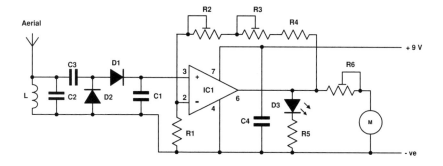

Figure 7.6 *Transmitter detector with LED and strength meter*

the gain of the amplifier. To set up the circuit, the variable resistor R2, the 'coarse' control, is set to maximum, giving maximum gain from the op-amp. It will be noted that the led may falsely trigger. This control is now turned slowly, until the false triggering ceases. Using the fine controller, R3, adjust until the led is just on the verge of turning on. The unit is now at maximum sensitivity. If the RF tuning capacitor, C2, is omitted, the detector will be more sensitive to frequencies up to, and including, UHF.

Component listing for the transmitter detector with LED and signal strength meter, Figure 7.6

Resistors

R1	470 R	R4	22 k
R2	1 M pot	R5	470 R
R3	47 k pot	R6	22 k pot

Capacitors

| C1 | 1 nF | C3 | 330 pF |
| C2 | 5 pF | C4 | 100 nF |

Semiconductor
IC1   CA3140

Meter   optional, 1 mA

Inductor   see text

A transmitter detector with both an LED visual display and an audio warning is shown in Figures 7.7a and 7.7b. The first section, based around IC1, is similar in operation to the above design. The second section, based around IC2, is a voltage controlled oscillator. To set up

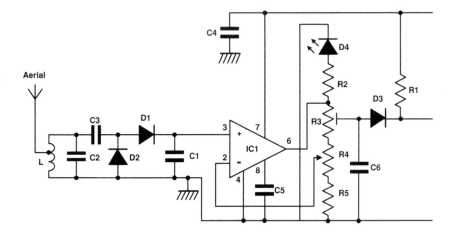

Figure 7.7a   *Transmitter detector with LED and audio indicator*

the circuit, temporarily connect a 1 M resistor between pin 3 of IC1 and the positive supply line. Slowly rotate variable resistor R3 until the pitch of the audio signal ceases to rise any further, then back it off slightly. Now it is necessary to adjust variable resistor R8 until the audio signal output pitch is at the highest frequency required. If the 1 M resistor is removed, it should be noted that the frequency of the output would drop to the lowest value of around 100 Hz. Now

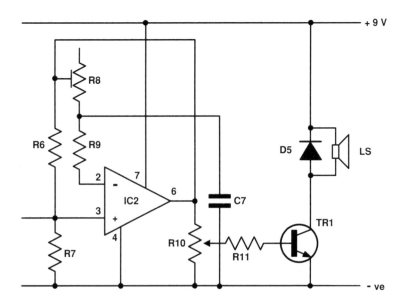

Figure 7.7b

rotate the sensitivity control, R4, to maximum, and the volume control, R10, to a comfortable level. Whenever the detector is within an RF field, the LED will illuminate, and the pitch of the audio will increase. Whenever a detector is used, it is imperative that any area is swept with the telescopic aerial of the detector in the same plane as the aerial of the transmitter, i.e. the aerials must be both vertical or both horizontal, since the signal may not be detected if aerial polarizations are opposite.

Component listing for the transmitter locator with visual and audible indicator, Figures 7.7a and b

Resistors

R1	1 M5	R6	100 k
R2	470 R	R7	100 k
R3	10 k pot	R8	250 k pot
R4	10 k pot	R9	15 k
R5	47 R	R10	100 k pot with switch
		R11	15 k

Capacitors

C1	1 nF	C4	100 nF
C2	5 pF	C5	10 μF
C3	330 pF	C6	100 nF
		C7	22 nF

Semiconductors

IC1, 2	CA3140
TR1	BC182
D1, 2	OA91
D3, 5	1N4148
D4	LED

Loudspeaker	64R

Inductor	6 turns, 5 mm dia., 22 swg tinned copper wire, tap 2 turns from earth end for aerial

Telescopic whip aerial longest possible

Note that some companies have from time to time advertised 'bug detectors' that will operate a silent vibrator if an RF field is detected. The vibrator can be placed in a pocket, or held in the hand, giving a silent warning that all may not be well. It would be possible to convert the above circuit to perform this task, by substituting the LED and current limiter resistor, R2, with a buzzer (with the sounder removed), so long as a back emf protection diode is connected in parallel with the buzzer, and current requirements are observed.

Although these RF field detectors can warn of a signal, they do not give an indication of the contents of the signal. The radio signal may originate from an illegally planted radio transmitter, but it may be just as likely that the signal does actually originate from a totally legitimate source. This source may be a commercial music radio station, law enforcement or other public utility transmission. The only way to confirm that the radio signal is from a clandestine, hidden transmitter, is actually to listen to the unmodulated information. This would mean using a receiver, and since it is too costly to build a suitable receiver, a ready made unit is preferable. A useful device to complement a radio receiver is a hand-held frequency counter with a telescopic pick-up aerial. These devices are available with excellent sensitivity, with full control over the same. A frequency counter will display the frequency of a nearby weak transmitter (or the frequency of a distant powerful transmitter). After sweeping an area with a frequency meter, it is possible to program a scanner receiver with a list of all suspect transmission frequencies, so that they may be monitored, and then eliminated from enquiries.

It is important to note that since a hidden transmitter may well be voice or sound activated, it may be necessary, during a search, to introduce an easily recognizable sound into the area, in order to activate the transmitter. Remembering that there are also 'silent carrier' transmitters around, any silent carrier must be investigated. However, all scanner receivers generate several 'birdies'. These signals are produced by the receivers' own circuitry, and appear on various frequencies, and should not be confused with sub-carrier receptions. Any scanner manufacturer will include a list of birdies associated with each individual receiver.

Testing telephone wiring systems

Telephone wiring systems may be tested to try and find out if a transmitter or tape recorder switch has been attached. A simple series of measurements may be used to detect a high proportion of devices as follows.

Note that typical voltages for on-hook and off-hook modes will differ slightly from the values stated in the examples, and will depend on several variables such as the telephone handset used, different telephone systems used in a different country, etc.

Voltage measurements

1 With the telephone connected to the system, the voltage across the supply lines is measured. This voltage should be around 48 V with the telephone handset on the hook, which should drop to around 12 V whenever the handset is lifted up off the hook.

2 If a series device were connected to the line, the on-hook voltage will still be 48 V, but the off-hook voltage will rise to a voltage that is greater than the normal voltage. The increase in voltage is caused by the resistance in the circuit being increased, with a corresponding decrease in the amount of current being drawn from the telephone system. In the example given in Figure 7.8, the reading is for a telephone with a 330 ohm resistor (the series transmitter shown elsewhere in this book), in series with the telephone handset. Note, however, if the resistance of the series device were much lower, then the increase in potential difference would be smaller than that in the example.

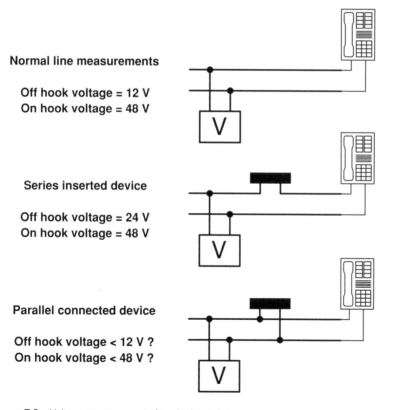

Normal line measurements

Off hook voltage = 12 V
On hook voltage = 48 V

Series inserted device

Off hook voltage = 24 V
On hook voltage = 48 V

Parallel connected device

Off hook voltage < 12 V ?
On hook voltage < 48 V ?

Figure 7.8 *Voltage measurement of a telephone line*

3 If a parallel device were connected to the telephone line, then depending upon the design, the voltage both on and off the hook could be found to be lower than the normal values, but a more elaborate test, using a device such as a line impedance tester, should be carried out. An audio frequency signal is fed down the line, which has been shorted out at the far end. A load resistor is placed in series with the line, with an oscilloscope connected across the load resistor. The frequency of the audio generator is swept from around a few hertz up to several kilohertz, with the voltage across the load resistor being measured on the oscilloscope. Depending on certain factors, such as the design of the parallel device, and the position of the device on the line (if connected very close to the shorting wire end, results are poor), it should be possible, when the injected audio frequency matches that of the resonant frequency of the audio pick-off components of the parallel device, to note a difference in the voltage drop across the load resistor, as indicated on the oscilloscope. Many types of test equipment could be used such as an impedance bridge, or a circuit built around an op-amp device, such as that shown in Figure 7.9, may be implemented in a bid to spot an elusive parallel device.

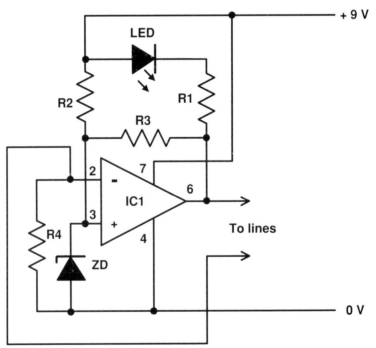

Figure 7.9   *Op-amp line tester*

Component listing for the simple op-amp line-checker, Figure 7.9

Resistors

R1	470 R	R3	3 k3
R2	3 k3	R4	10–33 M

Semiconductors

LED	light emitting diode, red
IC1	741 op-amp
ZD	3 V3 zener diode

Resistance measurements

With reference to Figure 7.10:

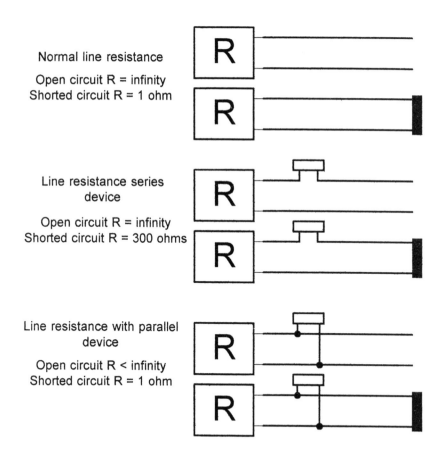

Figure 7.10   *Resistance tests on a telephone line*

1 Using a digital voltmeter, electricians 'Megger' tester, etc. with no equipment or socket components on-line (please note the warning below), the line should be tested for resistance, which should be infinity (no connection between the lines except perhaps for a very, very small amount of leakage through the wire insulation on very, very, long runs). With the ends of the line pair shorted together at the furthest point, the resistance should be no more than 1 ohm, depending on the resistivity of the wire used in the system.

2 If a series device is inserted in the line, then if the lines are shorted at one end, then the resistance of the series device will be measured, in the example shown, the meter would indicate a reading of 300 ohms.

3 If a parallel device was on-line, the test with the telephone pair open circuit may show less than infinity, as the very high impedance of a parallel may be 10 or 20 Megohms and hopefully be detected.

A word of warning: if an insulation tester is used, of either the cranking handle or push-button inverter type, it can generate between 500 V to 1 000 V. This can damage telephone equipment (not only in the building, but at the exchange), as well as some attached, non-authorized devices. Also, spare a thought for any tele-communications engineer, several metres up a telephone pole, who may not appreciate unofficial line checks. Whilst you are happily cranking the handle of your tester, the engineer may be in agony.

Other types of detection equipment

Hand-held metal detectors, the kind sold for cable and wall stud tracing at DIY stores, may be used to search for hidden microphones in walls, ceilings and floors. Wooden floors are excellent for hiding miniature microphones, since metal detectors will often believe the microphone element to be a floorboard nail.

Tape recorders could often be detected, since they used a system that utilized an oscillator of around 100 kHz. This could be picked up using a search coil comprising a few hundred turns of wire in a loop, feeding a detector circuit. Nowadays, d.c. biasing, which is undetectable, is very common in tape recorders.

When searching for electronic devices that contain semiconductor devices, such as bipolar transistors, a unit known as a non-linear junction detector can be used. These devices are used in store exits where the posts on either side of the doors are detectors, and the stock item tag contains a very simple diode circuit. The detector radiates a strong RF field, and when a non-linear junction is in close proximity, the latter will re-radiate harmonics of the fundamental frequency, which are in turn picked up by the detector. This device will find transmitters that are not even transmitting, even devices that have dead batteries. The detector will not work with transmitters that are built from field effect transistors, or transmitters that are shielded except for microphone and aerial outlets.

Telephone conversations may be scrambled, but for the average target, this may be too costly if conversations are believed to be of a non-critical nature. If it is believed that conversations are being recorded by a non-professional operative, a simple way of masking a telephone conversation is to connect an audio signal generator to the line. The telephone system accepts that a bandwidth of around 3 kHz is enough to transmit intelligible speech. An average tape recorder will have a response of around 10 kHz or higher if intended for high-fidelity music use. If the signal from the audio frequency generator is set to around 10 kHz, and is of the correct amplitude, it should be possible to carry out the conversation, but a tape recorder will become swamped with an irritating high frequency whistle. The only way to remove the whistle would be to process the signal with a notch filter or low pass filter, which may prove to be more trouble than it is worth to the amateur eavesdropper.

Snooper scarers

Although the largest portion of this book has been dedicated to the reception (or interception) of audio information or video input, a good deal of information can be obtained from a particular situation, whereby if an incident occurs, then the operative will be fully aware of an event, whereas the target would be completely oblivious to the fact that they have been 'rumbled'.

Snooper scarers are extremely useful in the workplace or at home if it is required to ward of the opportunist, without having to resort to the purchase or hire of a full-blown intruder alarm system or unnecessarily expensive surveillance equipment. A snooper is an unsavoury individual and is the sort of person that might hunt around for information in an office during lunch breaks, or after hours, looking through paperwork in in-trays or on desks, through drawers or cupboards for something to steal, such as a personal stapling machine, or gain secrets that may be used in the day to day political struggles that occur in every establishment. Since it is not possible always to alarm an office area due to cleaners or delivery personnel moving in and out, or maybe access through the office is required at all times, this is also the time when the part-time, unofficial 'investigator' strikes, or may hang back during coffee breaks to do their 'thing'.

In domestic situations, it may be necessary occasionally to temporarily alarm a room to prevent visiting small children wandering into a dangerous situation – maybe a pool area or medicines drawer in a bedside table. Also it is possible that someone will occasionally gain access to certain areas of the home where they have no right to be. Somewhere that cannot, for one reason or another, be locked or secured in any fashion will benefit from one of these devices.

The problems mentioned above may perhaps be addressed by using one of two methods of attack. The first method is to have an alarm that will give a loud, ear-splitting output if the input device is activated. This will have the effect of warning off the snooper, or letting the snooper become aware that you suspect something. The other method is to have a latched alarm that is silent, which will either give a local or remote indication that the input device of the alarm circuit has been triggered. This second method of protection means that the snooper will carry out their devious duties unaware that they have been detected.

The idea of a silent alarm is often used in alarm systems in conjunction with panic alerts. This means that the alarm is indicated at a remote or central station without the intruder being aware that they have triggered the alarm, which gives the security authorities time to arrive at the scene before the villain has fled the scene of the crime.

A design for a simple silent snooper detector with a latching action is shown in Figure 7.11. This design has an input for normally closed detector switches, which, if activated, cause the base of the transistor to rise and therefore switch on the transistor. The relay will close and since one pair of the contacts are in parallel with the transistor, even when the transistor is switched off by closing the switch, the relay will still be activated. The other pair of contacts on the relay are connected to an LED in series with a current limiting resistor. With this particular circuit, the relay will stay latched until power is cut off, perhaps by means of an hidden keyswitch.

It is up to the individual to decide on the type of detector switch used and it may be preferred to fabricate one from springy pieces of brass, etc. that can be activated when papers or folders are picked up off the top of an office desk, or removed from a drawer. The type of 'pressure switch' used in musical greetings cards can sometimes be used. The biggest drawback to fitting snooper alarms is doing so at a time so as not to draw attention to any would-be intruder.

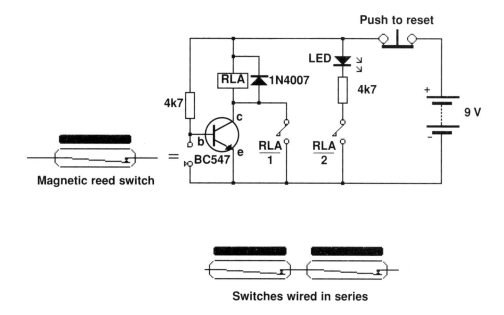

Figure 7.11 *Snooper alert silent indicator*

Telephone line in use indicator

Sometimes it is necessary for a person to know if any telephone handset on a particular line is being used, i.e. off the hook. In this situation a telephone line in use indicator is used so that the length of time someone is illegally using a phone line can be observed. The indicator is also beneficial to determine whether a piece of clandestine apparatus can be safely attached to a line without giving the target cause for concern by the installer causing a tell-tale intermittent line as the work is carried out, or when it is necessary to upload or download information on the phone line without corruption if a second party can readily observe that the line is already in use.

Figure 7.12 shows an indicator that is connected to a telephone wire pair so that a red LED is illuminated whenever a telephone handset is taken off the hook. The circuit is polarity sensitive, so trial and error must be used to find out which way round that the two phone line wires must be connected. The unit can be built into a case, with a switch in series with the battery to conserve power if so required, and a flying lead terminated with a telephone plug suitable for the country in which the device is used.

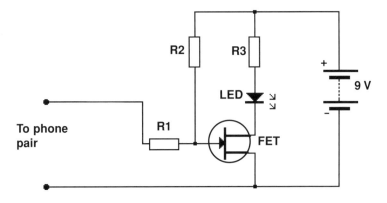

Component listing for telephone in use monitor

R1 = 10M   LED = red LED
R2 = 3M3   FET = 2N3819
R3 = 2k7

Figure 7.12   *Telephone in use monitor*

Simple polarity changeover switch

It is useful to have some way of quickly switching the polarity of two telephone lines since some circuits, such as the one described above, are polarity sensitive. The quickest and cheapest method of performing this function is to obtain a double ganged two-way toggle switch and to connect it up as shown in Figure 7.13.

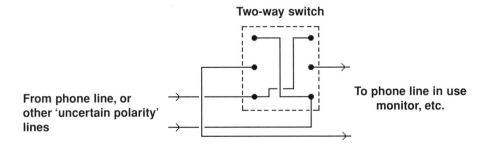

Figure 7.13   *Method of selecting correct polarity of undetermined polarity lines using a simple switch*

8 Receiving equipment

There is a large range of receiving equipment that can be utilized for electronic surveillance countermeasures. Scanning type receivers may be used to sweep the radio frequency spectrum for nearby low powered transmitters or even to detect the transmissions from higher powered covert transmitters from a longer range. Aerials may range from a simple telescopic whip type that can be used to trace a hidden transmitter in a room, up to a handheld or mechanically rotated multi-element beam aerial.

Scanning receivers

There is an excellent range of scanners available on the market today. Since the counter surveillance operative needs to find the transmissions of a device whose output is very low powered, in comparison to a commercial broadcast, it is a requirement that the receiver chosen has a very high sensitivity and very good selectivity. Since the frequency of the clandestine transmission is not known, it could, in theory, be anywhere in the radio frequency spectrum. For this reason, a very fast scanning rate is sought, so that the maximum frequency range is covered in the minimum time possible.

Every scanning receiver will have a certain number of what are known as 'birdies'. Due to the complex digital and analogue circuitry that is used in a scanner, there are certain frequencies produced by the circuits that fall within the reception coverage of the receiver. Whenever purchasing a receiver from new, the handbook that accompanies the unit will give a full listing of these birdie frequencies, which will vary from one specific model to another. Birdies will soon become apparent when the user scans throughout the range of the receiver and comes upon what seems to be a silent carrier, i.e. all that is heard is a carrier with no modulation or maybe with just a slight amount of hissing. If the aerial of the receiver is removed, and even if the aerial socket is short-circuited, the birdie will still be there. This simple test, along with a full list of birdie frequencies close to hand, will save time and trouble in trying to locate a non-existent transmitter.

Scanning receiver features

Although many scanners on today's market have several features that may or may not serve any practical purpose with respect to searching for transmissions from hidden transmitters, the basic features found on many scanners are listed below, along with an explanation of the terms used within this text.

Squelch

Every scanner receiver is fitted with a squelch control. When the control is turned anti-clockwise, any signal present, or white noise if no signal is present, will be heard through the speaker of the receiver, and the scanner will refuse to scan. When the squelch is turned fully clockwise, no white noise will be heard, only signals that are strong enough to break the threshold will be heard, and if in the correct mode, the receiver will begin scanning if the received signal ceases. This means that when checking for weak transmissions, the squelch control must be set so as to be on maximum sensitivity.

Limit

The scanner may well be fitted with a scan frequency limit. A section of the band is defined for a lower limit and an upper limit. This will mean that the scanner will start scanning at one end of the limit and when it reaches the other limit, will start scanning the portion all over again. This is a feature that means it is possible to sweep a section of the band many times over to ensure that the section is free from transmissions.

Memory

All scanners have a memory, ranging from around ten monitor channels to one thousand or more memory channels. It is therefore possible to enter in the frequency of a suspect transmission by direct keyboard entry, or to enter any interesting transmissions that have been heard into a monitor channel by the push of a single button.

Frequency synthesis

In the past, and still occasionally available today, simple scanners were produced that only had a very small number of channels. These multi-channel receivers were only capable of working on, say, a dozen channels, with one or more quartz crystals being required for each different frequency. With the design of the frequency synthesis system, which uses a digital process, a circuit in a scanner is capable of producing a vast range of frequencies which are used for reception, all of which are very accurate to a few parts per million. A crystal controlled receiver system, capable of receiving just one or two specific frequencies, are often marketed as part of a dedicated surveillance system, along with a matching crystal controlled min-iature transmitter, with an operating frequency typically on VHF or UHF. Apart from this particular use, a quartz crystal, or 'rock-bound' receiver, as they are very often called, is of no use whatsoever if wideband scanning is required.

Wideband scanners

This term can sometimes confuse the prospective purchaser of a scanner, since the terms 'wideband' and 'narrowband' FM are often used in the sales brochures. The term 'wideband scanner' refers to the ability of the scanner to cover a very wide band of frequencies. Receivers that only cover sections of the standard frequency spectrum are sometimes referred to as 'banded' scanners, i.e. they only cover certain bands. Since these second type of scanners have gaps in their coverage, it is often the case that the manufacturers of professional surveillance transmitters will follow the trend, and ensure that the output frequency of the product will fall into the gaps of the receiver!

The radio frequency spectrum

The term 'standard radio frequency spectrum' has been used in this chapter to describe the frequency range of typical transmitters that are used in electronic surveillance. As mentioned in a previous chapter, the frequency of a surveillance transmitter will generally be within certain bands, but may also be on the frequency that is least sus-pected. Some wideband scanners will cover a good proportion of the spectrum, starting off from the 'medium wave' frequency of 1 MHz, through the 'short wavebands' then up to 1 300 MHz or even higher, without gaps.

In Table 8.1 it can be seen how large the radio frequency spectrum is, if one considers the list as incomplete, since we are only concerned with the parts of the spectrum that are used in common surveillance transmitters.

Table 8.1   *Radio frequency spectrum*

Frequency division	Frequency range
Medium frequency (MF)	300 kHz–3 000 kHz
High frequency (HF)	3 Mhz–30 MHz
Very high frequency (VHF)	30 MHz–300 MHz
Ultra high frequency (UHF)	300 MHz–3 000 MHz

Typical surveillance transmitters may use a section between 70 MHz to 500 MHz

Modes of reception

The modes of reception that are available on the top-of-the-range scanning receivers are AM, amplitude modulation, FM, frequency modulation, plus ssb, single sideband, and c.w. (continuous wave). For use as a surveillance receiver, to detect voice modulated signals, only the FM mode would ever be used. Due to the relative complexity of the circuits required, no known surveillance transmitter using ssb has been produced. In a small number of applications, a transmitter that generates an unmodulated continuous wave can be used as a tracking transmitter on a vehicle, in which case the addition of a circuit, to give a signal that will beat with the silent carrier, or another method, so as to produce an audible tone from the speaker of the receiver, would be required.

With regards to AM and FM, many receivers do not have the capability of switching to FM whilst scanning the so-called airband, therefore, if an FM signal is picked up by the receiver whilst it is still switched to AM, the resulting audio will appear weak and distorted. The channel spacing of a scanner receiver is important. Several receivers will scan in 25 kHz steps in some bands, rather than the 12.5 kHz steps. These channel spacings are provided in a receiver because they are standard spacings for commercial purposes (10 kHz channel spacing is also popular). Since the frequency of a transmission from a covert device can be on a frequency that sits between standard transmission steps, it is preferable to obtain a scanner that can be switched from 25 kHz steps to 12.5 kHz steps or smaller. The smaller the size of steps, the longer the time needed to scan through the bands, but this will have the advantage of preventing any accidental skimming passed a weak frequency. It is therefore desirable to be able to go from wideband FM to narrow band FM to AM, etc. by using a switch that is provided on some models.

Choice of scanning receivers

When it comes to choosing a suitable receiver, it is entirely up to the individual, depending upon the portion of the frequency range that needs to be covered, as well as the required mode of reception. Receivers can be obtained that cover just a single band, such as an airband AM only unit, up to a much more expensive unit that covers a very large percentage of the standard radio frequency spectrum, with the ability to be switched from AM to FM reception throughout. The governing factor must be the intended purpose of the receiver, as well the proportionally higher cost of more complex receivers.

Table 8.2 lists only a small selection of scanning receivers that may be found on the market at present, and is not intended as any sort of indication of the author's preferences. New models of scanners are appearing every few months, some with more gimmicks than others, with other manufacturers perhaps favouring a more rugged and durable model, etc. Since the following information is only intended as a rough guide, no details of selectivity, sensitivity, etc. are included. For the purist, this additional information will be supplied upon request from the suppliers or manufacturers of the relevant equipment.

Table 8.2 *Scanning receivers – manufacturers and specifications*

Manufacturer Model	Station
AOR	
AR-1500EX	
Handheld	
Frequency Range	500 kHz–1.3 GHz
Modes	all modes
Memories	1 000
Scan Speed	20/second
AR-2000	
Handheld	
Frequency Range	500 kHz–1.3 GHz
Modes	AM, WBFM, NBFM
Memories	1 000
Scan Speed	20/second
AR-3000A	
Base	
Frequency Range	100 kHz–2.026 GHz

continued

Manufacturer Model	Station
Modes	all modes
Memories	400
Scan Speed	50/second
COMMTEL COM 1300 Handheld	
Frequency Range	8 MHz–1.3 GHz
Modes	AM, WBFM, NBFM
Memories	1 000
Scan Speed	20/second
COM 205 Base	
Frequency Range	25–512 MHz, 780–1.3 GHz
Modes	AM, WBFM, NBFM
Memories	400
NETSET PRO-46 Handheld	
Frequency Range	68–88, 108–174, 406–512, 806–960 MHz
Modes	AM, FM
Memories	100
Scan Speed	14/second
REALISTIC PRO-43 Handheld	
Frequency Range	68–88, 118–174, 220–512, 806–999.9875 MHz
Modes	AM, FM
Memories	200
Scan Speed	25/second
PRO-2035 Base	
Frequency Range	25–520, 760–1.3 GHz
Modes	AM, WBFM, NBFM
Memories	1 000

continued

Manufacturer Model	Station
ICOM IC-R71000 Base	
Frequency Range	25 MHz–2 GHz
Modes	all modes
Memories	900
YAESU FRG-9600 Base	
Frequency Range	60–905 MHz
Modes	all modes
Memories	100
YUPITERU MVT-7000 Handheld	
Frequency Range	1 MHz–1.3 GHz
Modes	AM, WBFM, NBFM
Memories	200
Scan Speed	16/second
MVT-8000 Base	
Frequency Range	100 kHz–1.3 GHz
Modes	AM, WBFM, NBFM
Memories	200
Scan Speed	20/second

Note that the list in Table 8.2 is only a rough guide to make the reader aware of the very wide range of receivers available, and is not complete. Although many of the receivers in Table 8.2 are listed as a base station, some of them will allow mobile operation by including a 13.8 V d.c. power connection. The actual frequency coverage of certain models may vary, as some are intended for sale in Europe, some are intended for sale in the USA, etc. Some receivers also have small gaps in their coverage.

From a legal point of view it is important to note that in many countries, it is illegal to listen into certain transmissions, and to do so may incur the wrath of the authorities. The penalties, if caught, can be very severe. If you intend to use a radio receiver on frequencies about which you may have any doubt of the legality, refer to the radio regulatory authorities in that country, state, etc. before doing so. Just because a particular receiver has coverage of certain frequencies, or bands of frequencies, that contain transmissions that are illegal to listen to, does not mean that the manufacturers or suppliers of the equipment condone the illegal use of their equipment, nor do you have an automatic right to listen on the grounds that you have spent a lot of money on the receiver. A band that is banned in one country will perhaps not be banned in another. Also it should be noted that the equipment is sold to many people who use the equipment during the course of their entirely legitimate business. Ignorance of the law is no excuse, nor will the attitude of, 'If they sell them, then why can't you use them?' be any use in court!

Spectrum analysis

Some companies supply a system that will allow a visual display of the radio frequency spectrum, picked up by their own scanning receivers, to be placed on a VDU (visual display unit). These devices will show any frequency being picked up, some systems giving a time, date, mode of transmission, etc. The system allows for magnifying, or zooming in, onto either a small section of a band or even onto one particular frequency. The price of these units will mean that their use is limited to only the most dedicated person.

Aerials

Telescopic and helical aerials

The most basic aerials that are available are those that are supplied when a receiver is purchased. The supplied aerial will be either a telescopic aerial if the receiver is a base station, or a flexible helical 'rubber duck' type with a handheld receiver. For tracking down transmissions with the handheld receiver, either aerial can be used. The problem with using an aerial with a wideband receiver is that the length of an aerial should, in theory, be altered, so that it is resonant with the wavelength of the frequency that the receiver is tuned to. The nominal length of a receiving aerial is one quarter of a wavelength. If the aerial is not resonant, then the reception of the wanted frequency will not be at maximum. If the aerial is more resonant to a strong, unwanted frequency, despite the filtering that is built into a receiver, the strong transmission may well break through. A telescopic aerial is preferable to a helical type for the following reason. When referring to the manual supplied with a scanner which is accompanied by a telescopic whip aerial, the instructions will tell you to extend the aerial to the full length when receiving the lower frequencies, but to use only one section if monitoring the highest frequency section. If using a flexible helical aerial, obviously the aerial length cannot be altered. A telescopic aerial can be useful for roughly pinpointing a hidden transmitter, especially if a variable attenuator is placed in series between aerial and receiver, since it may be collapsed to the minimum size allowable.

Wideband aerials

Wideband scanning aerials that are designed for rooftop mounting, are a cluster of aerial elements, each one to act as resonant on certain frequencies. They are almost omnidirectional, with a 360 degree coverage, since adjacent elements in the nest will slightly affect the others. The most popular design of wideband aerial is called a discone aerial. Because this type of aerial is bulky, meant for mounting on a rooftop, they are not useful for pinpointing trans-mitters. If a long run of coaxial cable is used between the aerial and receiver, this should be of good quality since losses in the cable will be greater as the frequency increases. It is also prudent to pay attention to any connector used, using only the better quality type, and avoiding joints, especially if outdoors, if possible.

Directional aerials

Directional aerials, similar to the type used as a UHF television aerial, although tuned to a specific band, can be useful in tracking down transmission. A directional aerial can be fixed to the top of a vehicle, and rotated to find the direction of a transmission. A scaled-down version can be handheld, and along with a receiver combined with an RF attenuator and strength meter, can be a useful tool. The dis-advantage of directional aerials can be their size, since if used at lower frequencies, because the length of the elements are correspondingly longer than those used at UHF, the number of elements and hence the overall gain and directivity, must be kept low. Another disadvantage of the directional aerial is that it has a somewhat limited bandwidth, so it would be necessary to have a set of aerials if full coverage were required.

9 Self-bugging

In this chapter, we will not be discussing DIY bugging, but will actually be looking at some of the ways in which a normal citizen may very often be eavesdropped upon, without a telephone tap or micro-transmitter in sight, as they go about their normal day to day lives. The intention of this chapter is to make the reader become aware of the potential ways that they may be inadvertently giving away details of their business and private lives to any 'snoop' or 'spook' who may wish to target them. Everyone uses some kind of electronic communication, most of which are open to the possibility of eavesdropping by an unscrupulous party without the use of specialist or complex equipment.

Cellular radio telephone systems

The first communication method to be discussed is one that is forever receiving publicity in the tabloids, due to the inherent ease of interception of some non-encrypted systems: the cellular telephone.

The cellular telephone system works upon a network, or web, of cells that cover a large percentage of the country. Each cell is controlled by computers, and whenever a unit goes out of the range of one cell, the link will then be automatically transferred to a neighbouring cell base station, thereby maintaining continuous coverage with a minimum of 'drop-outs', or disconnections. All cellular telephone systems are connected to the public telephone system, but it is the radio link in the system that is most vulnerable to attack by a casual or dedicated eavesdropper.

Although steps are being taken to eliminate the illegal practice of eavesdropping on these radio links, it will continue until drastic measures are taken. The law in most countries makes the actual act of unauthorized eavesdropping on telephone conversations, or the unauthorized interception of radio transmissions, illegal. Another step taken to make things difficult for the eavesdropper is the ban on scanning receivers that can cover the frequencies used by cellular telephones. This may lead to a receiver, that is intended to be sold in different countries, containing a diode matrix that is wired in accordance to each country's 'band ban', so certain portions of the receiving range are electronically locked out. With these receivers, it is sometimes only necessary to insert or remove a diode from the band inhibit matrix to defeat the lock-out. Another method of defeating the system is the addition of an RF convertor, that can receive the illegal frequency band and then convert this into a frequency that can be covered by a legitimate receiver. The LNB input of some Ku1-band satellite television receiver systems may operate on the same frequency as a cellular telephone, which means that a few centimetres of wire, connected to the LNB socket, could produce a makeshift receiver. Note that if these methods are used, then the user will still be breaking the law. At the time of writing, a secure cellular telephone system (now overtaking the non-encoded analogue radio telephone system) that uses digital encryption is perhaps the only secure method of communication, but look what they said about the 'Titanic'!

Within the UK, the frequency allocation for the cellular radio telephone network band is split into three separate blocks, i.e. TACS, ETACS and GSM. The first two are of an analogue nature, with the third block being digital. The frequency spectrum for these three blocks are:

- TACS 890–905 MHz mobile transmit
          935–950 MHz base transmit
- ETACS 872–888 MHz mobile transmit
          917–933 MHz base transmit
- GSM   905–915 MHz mobile transmit
          950–960 MHz base transmit

The channels are full duplex NFM, in 25 kHz steps with 12.5 kHz offset.

Cordless telephones

Many households and business premises now have cordless tele-phones installed. The majority of these cordless telephones are ideal for someone who needs a telephone that is portable, albeit with a rather short range, with a maximum range of 200 m. In the UK, these first generation telephones, CT1, are analogue, with a second gen-eration system, extended range CT1, with a much greater range of up to 2 km. The standard cordless telephone system is NFM, and uses the channels listed in Table 9.1.

Table 9.1   *Standard channels for cordless telephones*

CT1

Channel no.	Base unit Transmit frequency	Portable unit Transmit frequency
1	1642.00 kHz	47.45625 MHz
2	1662.00 kHz	47.46875 MHz
3	1682.00 kHz	47.48125 MHz
4	1702.00 kHz	47.49375 MHz
5	1722.00 kHz	47.50625 MHz
6	1742.00 kHz	47.51875 MHz
7	1762.00 kHz	47.53125 MHz or 47.44375 MHz
8	1782.00 kHz	47.54375 MHz

Extended range CT1

Channel no.	Base unit Transmit frequency	Portable unit Transmit frequency
1	47.43125 MHz	77.5125 MHz
2	47.41875 MHz	77.5500 MHz

If an operative were to listen on the base unit transmit frequency, they would be able to overhear both sides of the telephone conversation, with the audio of the cordless telephone owner being degraded in comparison to the audio from the other party. Would the operative require expensive receiving apparatus to listen in on the conversation? Although the signal used by the cordless telephone system is frequency modulated, it is possible, by using a technique known as 'slope detection', to receive FM signals on an AM receiver. Slope detection simply means slightly off-tuning the received signal using the tuning knob of the receiver. Although the recovered audio output is rather distorted, this method can give acceptable results, and the results are generally better if the receiver is of a cheap type with wide i.f. (intermediate frequency) bandwidth, and if the signal being received is relatively strong. A quick glance at the frequency coverage of a medium wave receiver will sometimes indicate that the tuning range covered is up to 1700 kHz, which will accommodate the first four channels of the UK cordless telephone system. If the operative wished to cover the remaining channels, they would either obtain a receiver that covered the full range of channels, or open up the receiver and alter the tuning range of the receiver by 'sliding up' the tuning capacitor. If the receiver was coupled to a VOX-type tape recorder, some serious eavesdropping could take place.

If all of this seems far-fetched, the problem of eavesdropping on cordless telephones using common domestic receiving apparatus was first brought to the attention of the author when spotting an article in a newspaper. Apparently, a person accidentally found out that they could pick up their neighbours' cordless telephone conversations on their personal radio. The unfriendly neighbour then overheard the owner of the cordless telephone arrange some work 'on the side', and straight away the Department of Employment were informed!

By either using a communications receiver with AM/FM switching, or improving the aerial system such as fitting an external long wire aerial to the family stereo, then it is possible to pick up cordless telephone transmissions from a distance that far exceeds the original 200 m range, to around 1 km or more.

Wireless intercoms

Wireless intercommunication devices are used for many purposes, for example, baby listeners, or for communications systems within buildings. The two types of transmission medium are those which use the mains wiring to carry the information, and the type that is plugged into the mains to derive power, with the information then transmitted by radio to a portable receiver. The type of intercom that uses the mains wiring as a transmission medium has two drawbacks:

1 The information may be intercepted if an operative has purchased an identical unit. All they would have to do is simply plug in their unit to enable them to eavesdrop on conversations, etc. Also the information may not necessarily be contained within the wiring of one or two rooms, and may be picked up throughout the office complex, block of flats, etc.
2 Because these devices are superimposing a radio signal onto the mains wiring, if a portable receiver is brought into close proximity of the wiring, or is plugged into a mains socket outlet, if tuned to the correct frequency, or harmonic thereof, the signal will again be intercepted.

Radio transmitter type baby alarms are usually comprised of a mains powered transmitter unit that is plugged into a handy mains outlet, with the receiver being a battery powered portable unit that can be taken away from the transmitter up to some 50 m away. The harmonics from these devices can often fall on the commercial VHF band, and so be picked up by anyone in close proximity while numerous specialist receivers can cover the fundamental frequency at a much greater distance. This particular topic will not be dealt with further, but if the microphone area can pick up sounds from other rooms, then it may be prudent to swap over from the radio transmitter system to a hardwired system during certain times.

Citizens band and amateur radio

Although it may seem like stating the obvious, whenever using these modes of communication, it is imperative that security and personal safety must be remembered during conversations. For every 'radio-active' person, i.e. one who actually goes 'on the air', there may be a few hundred listeners. Although you may wish to think that every listener is a 'good buddy', or a good and true radio amateur, there are always a small number of renegades and cut-throat opportunists. Even in these enlightened times, discussions are still heard on the airwaves such as 'We are all going on holiday on Friday', 'I am up here in the mountain, miles away from help if I needed it, with my wife and all my radio equipment that costs a lot of money', or, 'I am in the garden shed where I leave my very expensive gear unattended when I go to the Ham Club between the hours of 1900 and 2300'.

You may cry out that you have never given your address out over the air, but how many times do we hear a 'talk-in' for an 'eyeball', where the radio operator gives out enough information for any person with local knowledge to pinpoint the dwelling, then use that information at a future date?

Several pages ago, the question of why anyone should single you out for a target was asked. In the above paragraphs, it is illustrated that you do not have to be anyone special, for while people are talking to people, or while information can be grabbed from the ether, it can be guaranteed that someone, somewhere, wishes to hear the conversation, whether the reason is for financial gain, or just for cheap thrills. As some people are willing to go to any lengths or expense to eavesdrop, the level of care an individual wishes to exercise to avoid eavesdropping is entirely up to them.

Think on

The purpose of this book is to educate and inform the general public and security personnel of the world of electronic surveillance and of the devices and techniques that have been used, but does not condone the use of the same. The field of electronics is a wonderful thing. If you have never assembled an electronic circuit, then obtain one of the hundreds of excellent books that are packed with everything from how to build flashing lights up to how to build your own computer. Electronics is a tool, and like all tools, should be used responsibly and with care.

Index